20TH CENTURY VISIONS OF OUR FANTASTIC FUTURE

FOLLIES of SCIENCE

by

Eric Dregni & Jonathan Dregni

speck press

denver

Speck Press, Denver, CO

Published by: speck press, speckpress.com

ISBN 1-933108-09-6, ISBN13 978-1-933108-09-4

Book layout and design by:
CPG, corvuspublishinggroup.com

Printed and Bound in China

Library of Congress Cataloging-in-Publication Data
Dregni, Eric, 1968-
  Follies of science : 20th century visions of our fantastic
future / by Eric Dregni and Jonathan Dregni.
     p. cm.
  Includes bibliographical references and index.
  ISBN-13: 978-1-933108-09-4 (pbk. : alk. paper)
  ISBN-10: 1-933108-09-6 (pbk. : alk. paper)
 1. Technological forecasting--History--20th century. 2.
Inventions--History--20th century. I. Dregni, Jonathan.
II. Title.
T174.D74 2006
609'.04--dc22
          2006010580
10 9 8 7 6 5 4 3 2 1

To all the imaginative scientists, inventors, and artists who, with a healthy dose of skepticism, dare to dream of a better world and a fantastic future. And to our children: May they grow strong from the compost of our visions.

Thanks to Scott Fares and his I. A. S. B. S. (Institute for Advanced Studies in Big Science); Ruthann Godollei and her Wiggly Science; PK McCarthy (of course); to the Earth for support; the Moon for companionship; and the Sun whose energy made all of this possible.

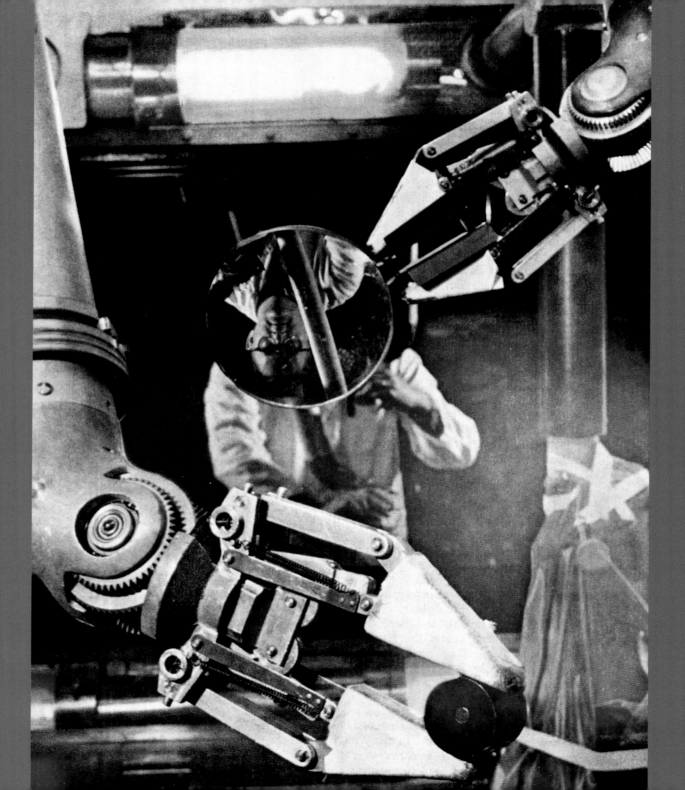

CONTENTS

ATOMS
FOR
PEACE

> "The future is called 'perhaps,' which is the only possible thing to call the future. And the only important thing is not to allow that to scare you."
>
> —Tennessee Williams,
> playwright of *Orpheus Descending*

Our generation is just as prone as any to futurist follies. The Dow Jones will reach 36,000; electric cars will cure pollution; the Cold War was the last war; hey, history itself came to an end in the '80s.

We all have visions of the future. The goal of science is to unveil ourselves of a prejudice and put forward instead a hypothesis. Maybe the Dow will hit 36,000, but how much of the prediction is rooted in hope, economic cheerleading, or greed? Curing pollution is a great goal, but even if cars don't belch emissions, aren't the millions of cars themselves, plus the millions of square miles of pavement, as much a part of the pollution problem? For Professor Fukuyama, famed political economist and controversial author of *The End of History and the Last Man*, to claim an end to history is one of two things. It is either a call to *1984*-style totalitarianism, or a view that everything of value has been bought and contained and nothing new can arise. Either way, it is frightening in its ignorance. Frightening, or laughable?

After the apex of the Industrial Revolution, a slew of new inventions emerged: Thomas Alva Edison's light bulb, Alexander Graham Bell's telephone, Karl Benz's internal combustion engine. With all of these incredible discoveries, the head of the U. S. Patent Office proposed in 1899 to close its doors because everything had been invented. Never mind

that Guglielmo Marconi's wireless radio had just begun to spawn spin-offs, Marie Curie was toying with radioactivity, Wilbur and Orville Wright would fly in 1903, and Albert Einstein was just getting going in 1905. Ol' Albert wasn't trying to change the world, just decipher it. "I never think of the future; it comes soon enough," Einstein said.

Ever since the Renaissance, humans have been obsessed with inventing the future. The brilliance of Leonardo da Vinci's inventions and ideas still baffle scientists and scholars. What would he invent if he were alive today and had all our modern technology? The fanciful minds of Jules Verne and H. G. Wells took inventions of their times and extrapolated into their futures. Climb into a hollow bullet and shoot it to the moon through an enormous rifle! The future seemed rosy and full of delightful Victorian gadgetry.

Then that brilliant curmudgeon Aldous Huxley shattered these utopian daydreams with *Brave New World* in 1932, set in the year 632 AF (After Ford). Art is banned, beauty is disruptive, and emotions are frowned upon except at the "Feelies" (multi-sensory movies). The masses are kept stupidly happy by a drug called "soma" (Prozac®?).

Even at the height of the Depression (economic, not psychological), America looked to the future for inspiration. Europe hadn't yet plunged into the slaughter of World

War II, so all eyes were turned to the fabulous Art Deco of the 1939 New York's World's Fair. General Motors' "Futurama" exhibit was a favorite and optimistically predicted the future:

"America in 1960 is full of a tanned and vigorous people who in twenty years have learned how to have fun.… When Americans of 1960 take their two-month vacations, they drive to the great parklands on giant express highways. A two-way skein consists of four, fifty-miles-per-hour lanes on each of the outer edges; two pairs of seventy-five-miles-per-hour lanes, and in the center, two lanes for 100-miles-per-hour express traffic. Strange? Fantastic? Remember this is the world of 1960!"

What happened to all that leisure time? Weren't we humans of the future supposed to laze by the pool with nothing to do except order another gin fizz from the robot waitress?

The boasted technology of the 1939 World's Fair did not channel us toward this leisurely future but toward a particularly devastating war, ending with two Japanese cities instantly obliterated. Still, kids' textbooks remained optimistic that we would prosper in spite of ourselves. "Through science, things that once seemed quite impossible have become not only possible but often commonplace. Just as this plane moves rapidly on toward new frontiers, so also does science," according to *The New World of Chemistry* of 1947.

Then another Brit, George Orwell, came along to splash water in our collective faces with his 1948 book, *1984*. Big Brother was watching through the television sets, and doublespeak "Newspeak" twisted words until unpatriotic speech was impossible. Dystopia once again smashed utopian dreams.

In looking back at past visions of the future, our goal is to put these futuristic follies and fantastic inventions into a context. We have the benefit of hindsight to chuckle at Ray Bradbury's disturbing suggestion that Walt Disney should have run for mayor in Los Angeles, making the city into Disneyland, and making the crowded streets walkable. We can laugh in amazement at L. Ron Hubbard's pronouncement that, "The creation of Dianetics is a milestone for man comparable to his discovery of fire and superior to his invention of the wheel and the arch."

Context is the great tool of us post-moderns. We look at history as a bunch of people living, much like us, but with different worldviews. We understand that their worldview was as important to their decisions as anything else because it colored everything in their lives. We relate to them because we see that our context flavors how we see the world.

The modernist of yesterday had no such vision. He (it was usually a "he") saw himself as the pinnacle of development, the progeny of lineage of great men who set the course of history. Two contexts present themselves: that of the civilized who were just like him, and that of the mysterious and unnecessary barbarians at the gate. He had few questions of the goodness of his goals, nor the inevitability of his quest; and the "other" was nothing but a museum piece.

We may have set aside this idiocy, yet the intimacy of context still eludes us. Modernism's pure optimism is alluring, but sometimes difficult for us to share. We are so attuned to context that we think ourselves outside of context, just as Fukuyama imagined himself at the end of history.

The role of the historian is to show not just the momentum of events, not just the possibilities at each juncture, but the full human perspective, the ideas and the emotions and the visions. It is perhaps the greatest folly of science to think that mere words and pictures can evoke a true sense of another time, with all the fogs of perception and the clarity of purpose of our forbears. It is those fears and hopes that we hope to portray and we ask you suspend belief and invoke the smiles of these earnest visionaries. Imagine the future that could have been if market forces had been ignored, if war had been outlawed, if practicality had been thrown out the window, and if we'd have obeyed Buckminster Fuller's advice: "Clearly we are here to use our minds." Maybe we'd now be driving his Dymaxion car on accident-proof roads, flying our jet packs to the pool in the domed city, and blasting off for a weekend trip to the rings of Saturn.

"The future is not what it used to be."

—Paul Valéry,
author of
*Introduction de la
Méthode de
Léonard da Vinci*

"The future, according to some scientists, will be exactly like the past, only far more expensive."

—John Sladek, author of *The New Apocrypha*

The Unimaginable: For all our scientific measurements and technological control there are still forces that defy our understanding (above). And they will be used against us. Not only does this floating monster have an enormous cranium, it has an extra set of huge glittering psychic-eyes.

Jet Packs, of Course: Off to work (left)!

Futurist Manifesto

TIME AND SPACE DIED YESTERDAY

F. T. Marinetti became the father of a new art movement when he published his "Manifesto of Futurism" in 1909. The Italian poet and painter railed against the softness and complacency that he believed characterized nineteenth-century Romanticism. Instead, he preached of the masculine power and vitality found at the root of recent innovations in physics, technology, and the arts. The amazing speed of the motor car, airplane, and modern city life offered humans a means of escaping the drab, pathetic lives they had lead for centuries. Tradition, liberalism, and sleepy immobility were out. They were to be violently replaced by innovation, "feverish insomnia," and the beauty of the machine. He adored the recently invented automobiles and wrote about caressing them. "We went up to the three snorting beasts, to lay amorous hands on their torrid breasts."

He knew that automobiles were a vision of the future and predicted correctly: "Soon machines will constitute an obedient proletariat of iron, steel, aluminum at the service of men."

Marinetti embraced the technology of the twentieth century, making it the center of his philosophy, politics, and personal style. He thought that to progress into the future, all things from the past must be violently rejected. "We want no part of it, the past … . Against practicality we futurists disdain the example and admonition of tradition in order to invent at any cost something new." He scorned Italy's obsession with the "smelly gangrene" of the old and proclaimed:

"We mean to free [Italy] from the countless museums that cover her like so many graveyards …" The futurist vision of modern cities with skyscrapers and multi-layered roads, skyways, and bridges was strikingly accurate, however.

That was just the beginning of the futurist proposals, however, as shown in this excerpt of Marinetti's "Futurist Manifesto" published in the Parisian newspaper *Le Figaro* in 1909:

> We want to sing the love of danger, energy void of fear.
>
> We confirm the world's magnificence has been enriched by a new beauty: the beauty of speed, the race car with its hood adorned with big tubes like explosive serpent's breath.
>
> We are on the extreme edge of the centuries! Why should we look back, when what we want is to break down the mysterious doors of the impossible? Time and space died yesterday. We want to live in the absolute after we've created the eternal velocity of the omnipresent.
>
> We want to destroy the museums, libraries, academies of every kind, we will fight moralism, feminism, every opportunistic or utilitarian cowardice.
>
> We want to glorify war—the world's only hygiene!

> "A preoccupation with the future not only prevents us from seeing the present as it is but often prompts us to rearrange the past."
>
> —Eric Hoffer,
> author of *The True Believer*

Obviously, by the end of his manifesto, he'd lost the support of most scientists who might have been excited by his technophile visions of the future. Marinetti's rants caught the ear of a young Benito Mussolini, however. The future Duce used the manifestos as the artistic foundation for Italian fascism. The unashamed exaggerations and celebrations of "the new" in Marinetti's writing turned-on some, and turned the stomachs of others.

Grand Futurists: Futurist Filippo Tommaso Marinetti at right, Bologna 1920 (above)

Life in the City: "City" (Cittá) by Italian futurist Virgioro Marehi, 1918 (above)

Where Will You Park Tomorrow?

"It may seem astounding that anyone would consider seriously the instantaneous sending, by any means whatsoever, of a solid object like a motorcycle, a set of dishes, or a man, from Paris to Chicago, for example. Yet possibilities have been analyzed, and some calculations made.... It won't be by 1990, but by 2090—or 2190—?"

—*Science of Tomorrow,* 1963

Heliport of the Future: As cars zip seamlessly by pedestrians (left), personal helicopters would land overhead. The city of the future (circa 1948) was still pedestrian friendly (and men still sported fedoras).

Gyropter Zeppelin!: "Features of an amphibian, a dirigible, a gyropter and an ordinary airplane," raved an article from 1930 about Guido Fallei's design (far right). One disk-shaped wing rotated on top like a helicopter while hydrogen, helium, or hot air added to the lift. In descent it just used a parachute. Yes, a parachute.

Tomorrow's Transportation
JET PACKS AND HOVERCRAFTS

SCIENCE IN THE SKY
ICARUS STRAPS ON HIS ATOMIC WINGS

EARLY AIR INCARNATIONS

Humans have always wanted to fly like the birds. Myths of winged people have been around as long as people have recounted them. Medieval inventors predicted air travel with human-powered wings, and in 1010, Eilmer of Malmsbury, England, reportedly flew more than 600 feet, at which time he broke both legs.

Leonardo da Vinci designed numerous flying mobiles that were a step in the right direction but never totally functional, relying on flapping but never on aerodynamics. Legend has it that in 1498, Giovanni Battista Danti flew one of da Vinci's human-powered ornithopters over a wedding party in his town piazza when one of the wings broke and he flew into the roof of St. Mary's Church.

Prior to Orville and Wilbur Wright's flight in 1903, hundreds of unsuccessful flying machines collided with *terra firma*. Immense creativity went into both the bizarre failures and the impractical successes afterward. The military tried many airborne ventures such as planes with folding wings to land on tight landing strips, and a saucer-like plane with four large ducts in the body of the plane, which pushed air down like a hovercraft.

Following World War II, one third of automobile dealers expected to be selling planes after the wartime *Stukas* and *Corsairs* captured the popular imagination. Many small towns built airstrips for the impending flight craze, and consumers eagerly awaited an aviating Henry Ford to build the "Model T" of the skies.

While planes never glided down to affordable prices, many two-in-one conversion kits tried to lure customers. The Fulton airphibian came out in 1946 to little success as a plane with a detachable cockpit that functioned as a car—"5-minute, 1-woman conversion time" from plane to car. The 1950 Moulton Taylor aerocar prototype hauled its wings on a trailer behind the car/cockpit for easy take off.

Rotavion: "The world's safest plane" (right below) in 1961, this half-helicopter could zip straight into the sky. Clearing the trees, it blasted forward until the aerodynamic lift of the wings took over. The air foils/Venetian blinds provided control, but more importantly, hidden jet engines could get the plane out of any trouble. Those jets can do anything.

Boston in the Future (right): Zeppelins, balloons, and bizarre Paul-Revere floating ponies vie for air space above Boston in this postcard from 1911.

An earlier, perhaps more practical idea, was the 1936 autogiro that blended a helicopter and an auto. One problem was the propeller mounted on the front grill, which would make quite a mess in any head-on collision.

As one of many aeronautic hybrids, the cleopter—part rocket, part plane—was designed by Helmut von Zborowski, who was associated with the company that built the first rocket aircraft, the *Messerschmitt*. *Popular Mechanics* complained of the 1958 blueprints, "One objection to the design: In the event that the engine failed, the crew would have no choice but to bail out."

AIRSHIPS AND ZEPPELINS

The first flying craze came in 1783 when Joseph and Etienne Montgolfier ascended into the clouds over France in a hot air balloon, although the brothers didn't surmise that it was actually the heat that raised them but thought it was an undiscovered gas.

The high point of ballooning came with the enormous stiff-framed Graf zeppelins. Predictions soared in the early 1900s as *Scientific American* predicted coast-to-coast highways of dirigible trains on rails. Even the Empire State Building's observation deck was originally built as a load-ing dock for blimps to board passengers and send them overseas to the Eiffel Tower.

Popular Mechanics ran a cover story on a combination zeppelin/ocean cruiser in the style of the *Titanic* and the *Hindenburg*. This ship could land on water for boarding and then fly over land or water for international flights. Unfortunately, the *Hindenburg*'s mid-air explosion, as well as many other zeppelin tragedies, made the risk too great.

"Look, Up in the Air, It's a Plane, It's a Kite, It's a Sailboat!": This sailboat of the sky (right) from 1930, was dragged to airspeed by an automobile. Once aloft, the jib and mainsail would be raised for added velocity. Unfortunately, any crosswind could tip the glider for lack of a centerboard or any real stabilizing force.

FRANK READE

WEEKLY MAGAZINE.

Containing Stories of Adventures on Land, Sea & in the Air.

Issued Weekly—By Subscription $2.50 per year. Application made for Second-Class Entry at N. Y. Post-Office.

No. 19.   NEW YORK, MARCH 6, 1903.   Price 5 Cents.

SIX WEEKS IN THE CLOUDS;
OR, FRANK READE, JR.'S AIR-SHIP THE "THUNDERBOLT."

BY "NONAME."

At this moment the Klamaths burst out of the cavern. But they had come just too late. Frank Reade, Jr., was in the pilot house, and the airship shot upward into the zenith.

"[The crew in an atomic plane] can't possibly be shielded from all radiation. The next part to the question involves a military judgement as to how much radiation you permit the crew to take in order to improve the performance of the plane."

—Professor Lyle E. Borst, Atomic Airplane Project

"Heavier than air flying machines are impossible."

—Lord Kelvin (1824–1907), Scientist and President of the British Royal Society

ATOMIC AIRPLANES

After the first atomic explosion, inventors envisioned nuclear reactors in everything from toasters to lawn mowers to rocket ships. The impatient public could hardly wait for an atomic airplane to be hovering overhead just a few years after the explosive end of WWII.

By 1956, five companies were working on prototypes, and the U. S. Navy predicted that the first atomic airplane would be a seaplane. Walt Disney pushed the concept in his "Tomorrowland" exhibit in Disneyland and in Heinz Haber's book *The Walt Disney Story of Our Friend the Atom*. He predicted that an atomic plane would fly as long as the crew wanted to stay in the air. "The atomic power reactor is encased in a heavy lead shield to protect the crew against dangerous radiations. Crew and passengers are positioned at a safe distance from the power plant.... The crew is further protected by plastic shielding and double-walled windows. The free space between the windows is filled with water, which absorbs any stray radiation from the reactor," according to *Our Friend the Atom*.

Popular Mechanics backed up Disney's story with an article in 1957, "The most desirable place for the crew is in the nose and, since the crew should be as far away from the reactor as possible, the nose may be exceptionally long."

Even the prudish *National Geographic* hopped on the nuclear-plane bandwagon with an article entitled "You and the Obedient Atom" in 1958. It bragged of scientific progress with human guinea pigs: "Lockheed...is looking to the future by testing the reactions of crewmen during long confinement in a simulated nuclear aircraft cockpit."

One of the problems of the A-plane was that all the lubricating oils in the landing gear and elsewhere on the

Lead Zeppelins: Packed full of the lightest-known gas, hydrogen, zeppelins (right) fell out of favor when the *Hindenburg* erupted into flames over New Jersey. The magnesium proved more flammable than wood. Helium is twice as heavy and more costly than hydrogen but ultimately safer. If this inert gas had been used in the first place, we'd be circling the globe in balloons rather than jets.

"It appears to me that the possibilities of the airplane have been exhausted."

Thomas Alva Edison
(1847–1931), American inventor

Catch a Hot Ride Behind an Atomic Plane: Circling Earth eighty times before refueling, the A-plane (far lower left) would pull gliders from its wings and tail. At the destinations, the gliders would simply detach and coast to Earth. Notice that the gliders would be bathed in the radiation plume of the A-plane's exhaust.

Silver Arrow Atomic Airplane (below): The only thing holding back this atomic airplane is landing and taking off—oh, and that pesky radiation leaking into the cabin. With cruising speeds of 10,000 miles per hour, the Silver Arrow, "…could take off from New York at noon and flash into Los Angeles at 9 a.m. the same day—three hours before it started!"

New York to California in MINUS 3 Hours

plane would turn to tar because of the reactor's radiation and intense heat. But of course the biggest problem with a nuclear reactor with wings is the possibility of an atomic explosion raining down over the countryside.

JET PACKS

During the "Jet Age" of the 1960s, people dreamed of strapping on jet packs and zooming into the sky like the Jetsons. These high-powered backpacks were a lawsuit waiting to happen; test runs sent passengers head over heels for lack of a rudder. Even

if they managed to get into the sky, what if the engine had problems?

Nevertheless, crash-test dummies were armed with jet packs and flung from trains in 1957, in hopes of simulating the desired speed. Initially, these tests were to see if an airplane pilot could use a jet pack rather than a parachute when ejecting from a troubled plane. The dummies piggy-backed on supersonic trains and *Popular Mechanics* reported that, "Shoes and helmets would be ripped from the dummies." Luckily, mannequins were used rather than humans since, "one bounced across the ground three times and then roared over the head of a photographer.... Observers said it flew like an airplane, but the high-speed movies revealed that it made a complete flip."

Not willing to throw in the towel, Bell Aerosystems developed a "sky taxi" in 1967, which could propel two people with a jet pack between them. Oddly, the hydrogen peroxide jets only allowed for twenty-one seconds of 600-pound thrust, making for a very short trip with a quick descent. Nicknamed the "two-man pogo," this jet pack was designed for moon exploration or for eye-in-the-sky news coverage (or perhaps a news event in itself as the lack of control would send it pogo-ing through the sky).

Bell Aerosystems tried again in 1968 with a "new jet belt" that strapped right on to the passenger's back. Fortunately, a buffer of intake air between the passenger and the turbojet

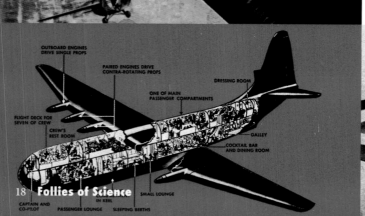

engine would prevent burning flesh, and a buffer between the fuel tank and the jet blast would prevent explosions.

SCIENCE SAILS THE SEAS
ATOMIC TIN CANS BENEATH THE WAVES

SUBMARINES

Submarines were scoffed at when they were originally conceived. A Yale college student named David Bushnell built the claustrophobic *Turtle* submarine, which housed one person who propelled the machine with a clumsy hand crank. The *Turtle* attacked a British warship in New York Harbor during the Revolutionary War but was no match for the big ship.

The *Nautilus* built by Robert Fulton in 1800 failed to find a market. Finally a somewhat successful submarine, the *Hunley*, was used by the Confederate Army during the Civil War. In Charleston Harbor, the *Hunley* rammed the Union ship, *Housatonic*, with explosives attached to a long pole. The pole wasn't long enough and the explosion sunk the submarine as well, making it a suicide mission.

The first engine-powered sub was invented by John Holland of New Jersey. It sank like a stone on its maiden trip in 1878.

Barrel Plane: The wasted space in airplane wings irked visionary designers (top far left). Why not get rid of the fuselage altogether and simply fly an enormous wing? With gigantic propellers pushing the air backwards through barrels, the tail-less plane would tumble head over heels, but physics be damned!

Royal Treatment: On the *Princess*, each transatlantic flight would be a royal ball (far bottom left). Rather than some surly flight attendant tossing a pack of peanuts on each fold-down tray, the *Princess* allowed passengers to don their cocktail dresses, sip martinis, and recline in luxury berths.

Dreams of Flying, Dreams of Falling: Decades after the Wright Brothers first flew, this French Air Force officer (center far left) wanted to prove he could fly with the birds. The wings were cut from plywood and his pants were webbed between the legs for stability. *C'est bon* that he carried a parachute to catch himself before plummeting to the earth.

Where's my Jet Pack?: Perhaps more than any other device, jet packs (far bottom right) represent the future. Finally, humanity would sprout wings and enter the Jet Age. "Possible uses: as a moon-exploration vehicle and for rescuing persons stranded in inaccessible places here on Earth, or for TV spot-news coverage." Never mind that steering a jet pack was deemed nearly impossible.

Look Ma, No Feet!: The inventor of the motorized skate, M. Constantini of Paris, once again struck gold (below) with his gasoline-powered swimming motor.

WATER LINE

but

Eventually, the kinks got worked out then came atomic power.

Atomic submarines originally ran ocean water directly over the cooling rods and pushed out radioactive seawater. Only when the Navy realized that this would leave a traceable trail behind the sub that would be easy to detect with a simple Geiger counter did they abandon the idea. One of the engineers on the project approached a Navy general about the danger of radioactivity, but his complaint fell on deaf ears. The general had a nameplate on his desk made of pure uranium.

The dangers of radioactivity were hugely underestimated, and water was considered an excellent radiation shield as it was so easy to purify. An article at the time claimed, "Drinking water, contaminated by radioactive fallout in atomic war, can be made safe in thirty minutes by a process basically the same as that used in home water softeners."

The U. S. Navy came up with another head-scratcher in the early 1950s to decontaminate its aircraft carriers. The simplistic technique involved water sprinklers that washed the entire ship with seawater to cleanse it of radioactivity from a nearby atom bomb blast.

RADIO ANTENNA
PERISCOPE
DERRICK STOWED
DERRICK IN OPERATING POSITION
CATWALK
CARGO HATCH
BRIDGE AND OFFICERS' QUARTERS
GALLEY
STORAGE
HOLD
REFRIGERATOR
CREW'S QUARTERS
FRESH WATER
SWITCHBOARD
STEERING GEAR
DIESEL MOTORS
MESS

Cigar-Shaped Ships: This concrete submarine/freighter (above) was destined for speeds of more than 75 knots, according to inventor Hal B. Hayes of Berkeley, California. Its streamlining would create 27 percent less drag than other ships of that time and allow it to carry 30 percent more cargo. Hayes built a mini-prototype propelled by two Ford V-8s that reached 20 miles per hour across San Francisco Bay in 1944.

Dudley Scootright: Built by forest rangers in Ontario (right) in 1949 to capture poachers along Lake Huron, the Ice-Water Scoot could skim across water, ice, and snow. Pretty good, eh?

I Can Ski for Miles and Miles: When this skier reached the bottom of the mountain, he simply flipped over his skis and revved the engine (above). The four tank treads brought him right back to the peak for another run. Each ski weighed thirty pounds, and the engine on his back added another forty-seven pounds, making for some spectacular wipe-outs on the slopes.

Ski-Don't: In 1901 when the first bicycling craze swept the country (far bottom left), cyclists didn't want pesky snow and ice to stand in their way. A little spiked wheel gave traction and propelled the two skis and courageous rider forward. Only cowards wear helmets, or worry that the designer overlooked any sort of braking device.

"Aerial Sedan:" With four large fans encased in air ducts (right) to prevent mutilating pedestrians, this 1957 hovercraft was projected to reach cruising speeds of 60 miles per hour. Hiller Helicopter was developing an "aerial jeep" for the military and realized the vast commercial possibilities of hovercrafts, if safe and reliable.

TRAVEL ON TERRA FIRMA
COAST-TO-COAST TELEPORTING

HOVERCRAFTS

Dreams of the future inevitably included hovercrafts riding on a cushion of air, flying a foot over the earth. Hiller Helicopter Company unveiled its plans for an "aerial sedan" in 1957. They hoped it would be on the market within a decade: "You'll be able to order a four-door model, a sports job, or even a light-truck configuration."

Ford Motor Company's Advanced Styling division took that prediction to heart with a 1956 model of the Ford *Volante Tri-Athodyne*, with two ducts in the back of the hovercraft and a single duct in front with dual fans. The fans would both lift the car and propel it to cruising speed as it traveled at a fifteen-degree angle with the nose aimed towards the ground.

A few years later, Ed "Big Daddy" Roth designed a personal hovercraft named *Rotar*, similar to the aerial sedan, but with only two fans powered by motorcycle engines in the rear to lift the mobile. Roth said that it "...would go

eighteen inches off the ground and forward at sixty-five miles an hour. But there are no brakes, and there is no steering on the three wheels, so once you came down on the ground, I think you might be in trouble." In fact, the last time it was demonstrated in the 1960s, the hovercraft exploded, hurting five people.

Nevertheless, the hype over hovercrafts continued, now with a new name: "ground-effect machines (GEM)." In 1963, futurist William Crouse expounded on the nuts and bolts of GEMs, saying they would be steered by air rudders over fields, rivers, or special roads that would be leveled, grassy fields. "Perhaps these strips can be put to productive use, raising low-growing vegetable or forage crops…. But perhaps by the time these highways are ready, new types of grass will have been developed that will not grow more than a couple of inches high and never need more attention." Crouse, however, lost credibility when he postulated that the "canals" on Mars are actually hovercraft routes.

TRAINS AND MONORAILS

Visions of futuristic public transportation raged during the economic downturns of the 1870s, 1920s, 1970s, and the gas-rationed war years. Ideas of elevated monorails zipping above cities excite people to this day, and the French TGV has made super fast trains ubiquitous in Europe.

The 1876 "Centennial City" World's Fair in Philadelphia sported a slow-moving Prismoidal Railway for Rapid Transit, straddling a wooden rail. This monorail was touted as the public transportation of the future, a claim repeated almost 100 years later at the Seattle World's Fair.

At the 1939 New York World's Fair, General Motors' "Futurama" exhibit envisioned the Big Apple in 1960 devoid of trains or any public transportation apart from the odd bus. Instead, seven-lane super highways were elevated to the tops of skyscrapers and cut through

Ford Volante: Ford would never forget when its *Model T* was left in the dust by GM's ever-changing models. It showed off its hovercrafts, the *Model Volante Tri-Athodyne* (above), and its "atomobile," the *Ford Nucleon*, years before the competition.

Rocket Train: Looking like a scene from Ray Bradbury's *Fahrenheit 451* (right), this 1938 rocket train would surpass 500 miles per hour and ride through rings of magnetized cobalt to keep it on course.

the middle of buildings and safely brought the future inhabitants to their garden plots in the suburbs.

This vision was realized in a unique way in Chicago. Wacker Drive, Upper Wacker Drive, and Lower Wacker Drive were created in 1926 as a subterranean road system beneath downtown Chicago. The drives are still in use today by Chicagoans experienced enough or drunk enough to dare visiting the underworld.

In 1955, General Electric predicted that trains would run on "levapads," a cushion of air a couple of thousandths of an inch thick, which would allow them to zoom at 200 to 400 miles per hour. The levapads were essentially an extension of the hovercraft dream with a track to allow some sort of steering capability. Of course ordinary coal, steam, or diesel would not provide enough power, so this was the dawn of, "a new era of railway electrification pow-ered by nuclear electric generating plants spaced along train lines, rather than individual atomic plants built into the locomotives." So far, no levitation trains have connected any cities, but a new high-speed maglev train, or magnetic levitation train, has been built in China with magnetic force between the vehicle and the guide rail to prevent any friction.

Ford picked up on levapads but knew that consumers wanted individual mobility, so it developed the individualized *Levacar Mach I* in 1959. Arguably the closest any auto company ever came to producing the Jetsons' car, the *Levacar* would run on special tracks and require little steering.

Super high-speed trains connecting metropoli inspired car designer Alex Tremulis to envision the gyronaut monorail armed with jets and gyroscopes to zoom rocket trains more than 200 miles per hour from town to town. While the rocket train with a stabilizing gyroscope may not have materialized, the Italian pendolino train uses a pendulum to keep it balanced at high speeds.

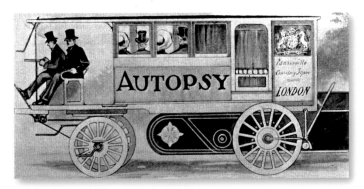

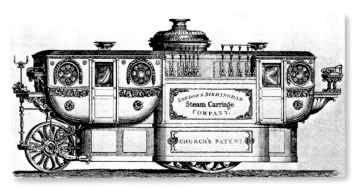

Autopsy—Death Ride 1836: Early self-propelled vehicles used sails or pedals, the first feasible power source was steam. The *Autopsy* (top left) was built by Walter Hancock in 1836, one of nine carriages for his commuter line in London, England.

London to Birmingham Steam Carriage: William Church ran luxury steam carriages (bottom left) from England's capital to Birmingham in 1832 and had plans to extend passenger service all over the country. Unfortunately for him, railroads soon ran him off the road with smoother and faster trips.

Rocky Road: Rather than just zoom ahead down the track, this gravity train (bottom left) took advantage of Newton's law that objects in motion tend to stay in motion. Enjoying a cup of tea in the cabin might prove tricky, however, as this was more of a perpetual motion roller coaster than a smooth streetcar.

***Mono Flyer* Moves (Past) Mountains:** Using the miracle of gyroscopes to balance on cables high above the valley (top left), the *Mono Flyer* was designed to flatten mountains as obstacles to transportation. Imagine the excitement if one of the engines stalled!

Treaded Bicycle: This patent drawing (bottom right) stems back to the days just after the Civil War when pedal power combined with a tank tread to create the original mountain bike.

Rocking Roller Blade: This torturous-looking device (far right) from 1903 didn't waste the downward energy of each step, but transferred it into forward motion and a spectacular balancing act.

Chain-Driven Roller Skate: By pumping down on the running board with each step (far left), skaters could reach amazing velocities, possibly without even running the gigantic skates from 1901 into each other. The front wheel was even armed with a rod brake in hopes of not tumbling head over heels.

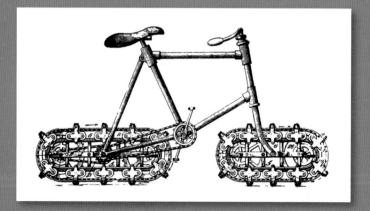

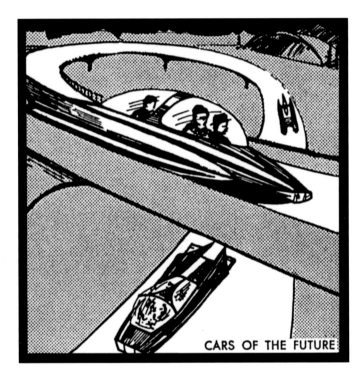

CARS OF THE FUTURE

Automated Autos: Cars were visualized with built-in radio receivers, communicating with the transmitters in the road, making an entirely automated trip and allowing the driver to snooze. The 1963 book *Science Marvels of Tomorrow* by William Crouse boasted about these automatic autos that would allow you to call, "...your friends on the car radiophone so that they are waiting at a convenient spot to meet you."

GYROSCOPES

Nothing screams "hidden powers" like the gyroscope. Uncanny in its ability to stand free of any supports, the gyroscope may be as close to antigravity as we'll ever come. And most of us have sat on one…in the form of a bicycle.

The gyroscope made giant was the dream of many inventors. Ideas ranged from huge, balancing cannons to enormous wheels ranging over land or over sea. The boomerang was the only useful gyroscope for millennia, before the bicycle.

But it is tiny gyroscopes that give airplanes and submarines the ability to keep their bellies down and their backs up, and without their frenetic spin, satellites would be unable to keep track of direction.

The most interesting use of the gyroscope came within the last decade, with the invention of the Segway. Like a two-wheeled stepladder, the segway uses computers to analyze gyroscopic balance and responds to the shifting weight of the rider to move. Coming soon: a one-wheeled motorcycle.

AUTOMOBILES

One of the few futuristic predictions that has come to fruition is the automobile. Once the ubiquitous "tin lizzie" became old hat, consumers demanded better roads for higher speed. Frank Lloyd Wright produced plans for a fictitious "Broadacre City" in 1935 that would level buildings to produce wall-to-wall highways. Sidewalks would have to be elevated to allow cars unlimited speeds.

While decent roads wouldn't become commonplace until the 1956 Federal Interstate Highway Act, the semblance of speed, in the form of streamlining, would have to do. Buckminster Fuller designed his space-age, three-wheeled dymaxion "omnimedium" wingless flight device (i. e., car)

Dream Car Built from Surplus

Homemade Dream Car: Rather than relying on bonehead designers in Detroit, why not just build your own space-age ride with scrap metal in your garage (above)? Before autos featured computerized ignition, do-it-yourselfers empowered with their ratchet set and acetylene torch made their own car according to easy-to-follow blueprints. Results may vary.

Flying Saucer Car: When WWII was in its final throes, the public was giddy to imagine future automobile possibilities. This dream car (far bottom right) put the engine in the rear and encircled the car with a bumper for maximum safety. One continuous window domed the car with little explanation of how the passengers would actually sneak inside this beautiful machine.

in 1934. He anticipated it could be fitted with jets under its inflatable wings that would fly like a duck (he used the word "plummet" often in describing the type of flight) traveling on land, air, or water.

Undoubtedly the biggest visionary of the automobile was Harley Earl and GM's Art and Color section, designing outrageous prototypes with futuristic features such as fins, airscoops, and ubiquitous chrome. Earl toured U. S. Air Force bases for inspiration, and in 1954, he designed the *Firebird 1* prototype. This model looked exactly like a small jet complete with gas turbine engine shooting out the back, a bubble dome as windshield, and an air-drag scoop that could slow the vehicle in case the brake pads on the wheels wore out.

Earl was obsessed with the future, so he began the "Motorama" autoshow in 1952 to show off his ground-breaking ideas. In 1955, he introduced Mr. and Mrs. Tomorrow with their own cars: hers, with "…its soft, sensuous shapes done in pastel colors," and his with "…angular lines and bold, brown coloring."

Auto engineers experimented with new materials for auto bodies, from aluminum to plastic on the 1954 Oldsmobile *Cutlass* prototype; to fiberglass on Corvettes; to stainless steel DeLoreans.

Once the basic design of the car was established in the minds of consumers, the frills were added. Everything from TVs to laser-beam guides to refrigerators on the 1969 Buick Century *Cruiser*, to radar control on the 1959 Cadillac *Cyclone* prototype were proposed. The dashboard was inevitably the destination for new gadgets, and one of the latest, a holographic display panel, was shown on the 1987 Pontiac *Pursuit* prototype. The 1951 Buick *LeSabre* prototype added an automatic jack control from the dashboard, while the 1953 Ford *XL-500*

very little, because it's such a nuisance to take them out. A waiting industry that will do wonders for prosperity will spring up when we revamp our cities and make it safe, convenient, pleasant, and easy to use a car on city streets." Refusing to own a car and using public transportation was tantamount to conspiracy against the system.

What's good for GM is good for the country, so the cities of the future would be ruled by the automobile. GM teamed up with Standard Oil, Philips Petroleum, Mack Trucks, and Firestone Tires to form a bus company, National City Lines, to help dismantle the clean streetcars. "Actually there is no harmful effects of the fumes and numerous tests have been made on that account," a GM spokesman said of bus exhaust. By 1946, National City Lines had converted eighty cities from streetcars to buses. In just eighteen months, New York's streetcars, the largest system in the world, was dismantled, and Los Angeles' trolleys were burned to make it the futuristic city of the automobile.

In 1953, President Dwight D. Eisenhower appointed GM President Charles Wilson as the Secretary of Defense, who immediately made wide interstates essential to national security. Ike agreed and declared in 1955, "Automobiles mean progress for our country, greater happiness, and greater standards of living."

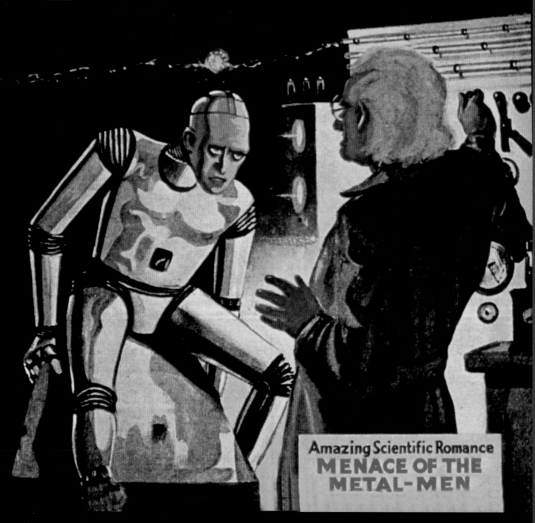

THE RED MAGICIAN: By John Russell Fearn

FANTASY
THRILLING SCIENCE FICTION

1/-

Amazing Scientific Romance
MENACE OF THE METAL-MEN

It's Alive!: What kind of mind would it take for a machine to walk? Why not a human mind? In this radio-controlled Frankenstein's monster (left), a human head was grafted to a spectacular robot body.

Computers and Robots
"SORRY DAVE, I CAN'T LET YOU DO THAT"

THE RISE OF THE MACHINES
BIRTH OF THE ROBOT

The word "robot" was coined in the 1920s, from *robota*, a Czech word meaning "hard labor." Before that time there were many *automata*, mostly toys or music boxes, which were hardly self-governing machines and did not perform hard labor. The first known automaton was from Greece, 400 B. C., an artificial pigeon that flew on an arm driven by steam. Watching *Indiana Jones* gives the impression that complex machines were part of the defenses of ancient Egyptian or Amazonian tombs and temples.

The age of machines took off in the early 1800s, with the invention of the cotton gin and the steam engine. England was peripheral to many of the conflicts on the continent, and after cornering the banking and mercantile markets it used its relative peace and prosperity to mechanize to a degree never before known.

During the Renaissance in Italy, very complex machines were built, mostly to handle agricultural chores like carding and weaving wool. Interestingly, they never sparked an industrial revolution, though Italy was at the time the center of finance and trade.

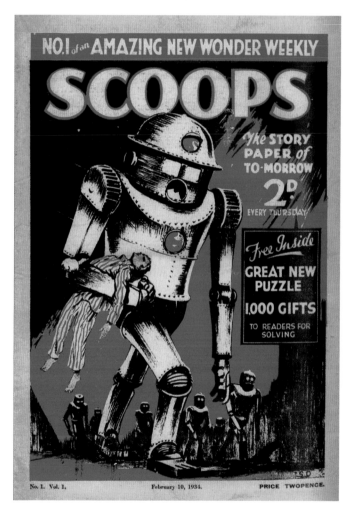

The March of the Pajama Stealing Robots: What horror these metal monsters exude! Formidable as tanks (above), they sowed destruction in the night, unstoppable and pitiless. Can't you hear the shrill metallic whine of their joints, the clank of their step, the amplified echo of their voices?

HARD LABOR FOR FUN AND PROFIT
INDUSTRIAL REVOLUTION ROBOTS

At the turn of the last century, small electric engines took over and workshops moved away from large central power sources, with their dangerous belts and oily shafts. The first true robots required the electric motor to allow them to be lightweight yet mobile.

Most of what we call robots are just complicated machines used in manufacturing. If you've ever seen socks made, for instance, you'd have a hard time explaining why the sock maker is not a robot, there are so many motions and patterns it handles. Robotic automakers hardly think for themselves. They might have some computers within that maintain quality control, but mostly we call them robots because they use superhuman strength to manipulate the car or engine as it's being assembled.

COMPUTE THIS!

The counting machines of old are fascinating gadgets of cogs and pulleys. Charles Babbage completed his "difference engine" in 1822 as a working model of a complex calculating machine, flawed only because of the coarseness of the materials available.

The abacus is a calculating machine that works much as a computer. You write a number by moving beads: You write the next number over the first, adjust for any digits carried to the next place, and a new number appears. It is very effective at handling huge strings of numbers, far quicker than a human punching numbers into a calculator. Most important, it teaches the user to think as a computer. Those proficient in the abacus can add, multiply,

Robotrix—Pure Human Beauty Enslaved in the Original Matrix: The ruler of Metropolis had this robot created as a double for a beautiful young woman (above). The robot had her own agenda: provoking riots among the workers so that robots could take over where humans so obviously failed. This dark dystopia featured videophones, stratoscrapers, and humankind dwarfed by the building-sized machinery they served, and prey to the machines they made in their own image.

1954 Ford *FX Atmos*: FX for "future experimental," and *Atmos* from "atmosphere." This space-age bubble top (below) had protruding aerials from the front for remote control of the vehicle by radio and "electrical control impulses," which warned of other autos. No need for bumpers since it would supposedly be impossible to collide with any obstacle. The radar screen was placed on the dashboard to view fellow spaceships and doubled as a television.

1969 *Astro III*: The trend of a car resembling a cockpit continued with the three-wheeled *Astro III* prototype (below center), complete with turbine engine. You would simply turn on to the automated highway of the future and the car-cum-plane would be zipped to its destination while the driver slept. No rearview mirrors were necessary; the pilot could look back through closed-circuit television.

prototype and 1965 Plymouth *V. I. P.* came with a handy dictaphone.

The 1961 Ford *Gyron* balanced on only two wheels with two retractable stabilizing or "landing" wheels for slow speeds. Balance was achieved by a gyroscope allowing the *Gyron* to lean into turns.

The 1969 Buick Century *Cruiser* could either run on automatic pilot or by manual operation with controls built into its armrests. The driver would insert punch cards telling the car of its destination and then cruise along the automated highways of the future.

ELECTRIC, ATOMIC, AND OTHER AUTOS

Since fossil fuels are a non-renewable and polluting source of energy, designers have constantly been searching for a more viable form of propulsion. In 1917, inventor John Andrews approached the U. S. Navy with a secret green powder that supposedly turned into fuel when mixed with water. The Navy wasn't convinced, and the mysterious powder went underground until 1973 when Guido Franch of Chicago again tried to sell the powder for a nominal fee of $10 million. It turned out to be a form of lighter fluid he claimed he got from aliens from Neptune, known as the "Black Eagles."

By 1907, twin brothers Freelan and Francis Stanley designed the rocket auto based on their earlier Stanley Steamers. The super aerodynamic steam car supposedly hit 132 miles per hour when it caught a bit of an updraft and left the ground. The rocket crashed to the ground 100 feet later and smashed the Stanleys' dreams of speed.

The 1920s and 1930s saw electric cars putzing around eastern cities, but they were limited by the battery to fifty

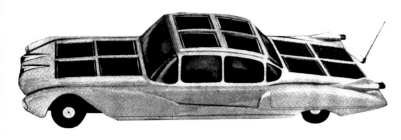

Solar-Powered Auto: While sunlight may seem like a viable alternative to non-renewable fossil fuels, this build-it-yourself kit was a little too simple and a little too flammable. "The car body is made of balsa wood in five sections, the two center sections being hollowed out to receive the pine chassis, motor, and drive mechanism," according to a review in 1957. GM, however, kept experimenting and developed the 1987 *Sunraycer* (above) solar auto, which won a 2,000-mile race across Australia. Students at the University of Minnesota also designed a super low rider solar auto in 2005 that zoomed along on sunny days.

Luxury Busses: According to GM's "Futurama" exhibit at the 1939 World's Fair, the busses of the future (below) would have, "...full-sized berths, bars, and observation 'decks' fitted with clear-vision plastic for luxury travel."

miles at a stretch. Because of this, they were considered an auto for women to do the shopping and other activities that would keep them close to home. Nevertheless, speculation on electric cars continued as in the 1963 book *Science Marvels of Tomorrow*, "A still different solution…would be to place electric induction coils under the highway. The car would pick up electric power, by induction, from these coils as it drove along the highway." Today, the Big Three are following Japan's lead with hybrid cars mixing both electric and gasoline power.

Another source of power was through a simple chemical reaction of oxygen and hydrogen displayed on the 1959 De Soto *Cella* prototype. Turbine engines were taken from planes and tested on cars as with the 1961 Chrysler *Turbine* car, of which fifty prototypes were produced and tested all over the country. It weighed hundreds of pounds less than regular engines and had far fewer moving parts.

Liquid fuel still remained the most practical form of energy, but the 1951 Buick *LeSabre* prototype added a second fuel tank inside its tail fins, using methanol for extra speed.

The most fascinating car of all was the atomobile or atomic car. While the nation was still reeling over the power of the A-bomb, *Popular Mechanics* ran an article in 1945 about an atomic car "with a 'butterfly' motor or U-235 engine, already well on its way to perfection, a car could be driven 5,000,000 miles without refueling." Testing continued until Chrysler attempted to halt speculation in 1957 saying that "a 3,000-pound automobile would require a nuclear-power plant weighing 80,000 pounds to move it along the highway."

Ford Motor Company thought otherwise and unveiled its model of a Ford *Nucleon* in 1958 with a mini-nuclear reactor in the rear trunk of the car. Although the actual engine was never tested, they produced a scale-model atomic car with a radioactive core suspended by booms on the rear of the car; they admitted that the shielding was not even close to enough to protect the passengers.

Both *Nucleon*'s axles were located toward the rear because of the immense weight of the reactor. Ford predicted that gas stations would eventually just sell small bits of uranium to refuel the *Nucleon* every 5,000 miles or so. The 1959 book *Atoms Today and Tomorrow* predicted even better mileage: "An atomic pill about the size of a marble has enough energy in it to drive an automobile four times around the earth."

"There will never be a bigger plane built."

—A Boeing engineer
after the first flight of the 247,
a twin-engine plane
that carried ten people

Pilots Needed: Multibladed helicopters (top right) landing on multitiered parking discs? Where do I get my pilot license! Note that the parking discs appear to rotate. Just another variable for these ace pilots.

Plane Meets Bus: Rather than tiresome loading and unloading busses to get to the plane on the tarmac, this pack plane (right) would simply attach the whole darn bus onto the nose of the plane and take off to its next destination.

City of Today: Mud and Manure

CITY OF TOMORROW: CARS AND CLEANLINESS

Cities in the 1800s stunk, literally. According to north St. Paul records, "Each horse would excrete fifteen to thirty pounds of manure each day, a lot of which was not removed from the city streets." In 1880 alone, New York City had to dispose of 15,000 dead horses that were often left to rot in the streets because of early gridlock during rush-hour traffic. Horses left something to be desired for urban transport and surely didn't fit into the city of the future.

In 1889, electric trolleys were zipping through the streets as pedestrians jumped out of the way of huge sparks from the tracks and the overhead lines. Passengers overcome with nausea from the bumps would chew gum to keep their lunch down and complained of the winter cold because the carriages were open to the elements. Nevertheless, ridership skyrocketed as manure decreased.

By 1905, nearly one-quarter of urban traffic was non-polluting bicycles. Only the wealthy could afford bikes, however, so working-class kids would often lob stones at cyclists. Opulent bicyclists made outrageous demands for smoother roads made of wood from endless northern forests, steel from the Iron Range, or macadam. The early bike craze soon faded as the horseless carriage was the new transportation craze.

"We shall solve the city problems by leaving the city," declared Henry Ford in his motorized march to colonize suburbia.

GM echoed Ford's plea for suburbanization in a 1924 Chevrolet ad: "The once poor laborer and mechanic now drives to the building operation or construction job in his own car. He is now a capitalist ... . Before or after acquiring the automobile he has begun paying for a suburban home of his own ... . How can Bolshevism flourish in a motorized country having a standard of living and thinking too high to permit the existence of an ignorant, narrow, peasant majority?"

Auto manufacturers began campaigning vigorously to remake cities to be car friendly. Like the president of Studebaker noted in 1939, "Chief MacDonald [of the Bureau of Public Roads] insists that we must dream of gashing our way ruthlessly through built-up sections of overcrowded cities in order to create traffic ways capable of carrying the traffic with safety, facility, and reasonable speed. The city of the future, in a word, is going to be vastly different from the city of today, swamped in the 'mud' of congestion." Frank Lloyd Wright agreed waxing poetic about Chicago's widened streets allowing cars to zoom through to suburbia. He envisioned elevating sidewalks above the streets so cars could look into storefronts. As a compromise, sidewalks were trimmed back to half their width.

The 1939 World's Fair in New York was awash in highway propaganda as GM presented a scale model of an Interregional Highway System including 40,000 miles of interstates. Their literature at the time declared, "If we are to have full use of automobiles, cities must be remade. The greatest automobile market today, the greatest untapped field of potential customers, is the large number of city people who refuse to own cars, or use the cars they have

The Runaway Robot

by LESTER DEL REY

"There is no reason anyone would want a computer in their home."

—Ken Olson,
President of Digital
Equipment Corporation

"We Have a Runner:" Technology on the loose, this image induces a sense of pity (top) for the enslaved machine. Run, robot, run like the wind!

"Forgive Them, Father, They Know Not…:" Note the ordered city in the background, the uniformed men approaching (above). This robot seems apologetic, hiding behind the destruction he has wrought. Is he truly a heartless beast, or is there redemption?

Unstoppable: Metal giants (right) with flashlight death rays, zapping planes from the skies like bugs.

JOHN RUSSELL FEARN

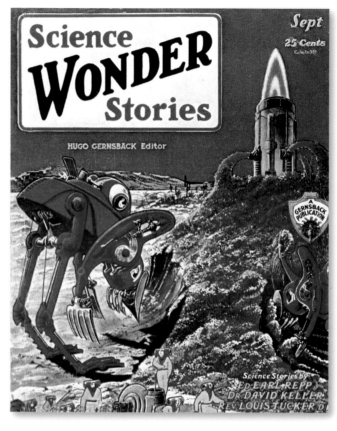

Mining the Blood of Earth: The age of machinery (above) and of remote control blends into the age of robots. At what point do huge machines, with superhuman strength (auto assemblers vs. ditch-diggers), become robots? Autonomous machines do not yet exist, though the Martian rovers are getting close.

THE DEFENDERS By Philip K. Dick

Robots Making Robots: We may have reached the point of no return when machines begin making themselves (above). Computers creating computers is the next point. Ah, but these robots look so intent, and so happy. Aren't we glad they at least found satisfaction in their work?

We Might Need Another Hero: The skies are on fire in the last days of humanity (left); even our prized fighter planes are being commandeered. These 'bots obviously mean business, but wait, isn't there a sense that a trap is about to be sprung, that some able-bodied hero will save the day?

and do long division in their heads, by simply imagining the beads moving.

The first thinking machine gained consciousness in 1942; during the war, punchcards handled the administrative details of the Nazi holocaust. Not a very auspicious beginning, this should have been grist for the mill of writers and technophobes. While modern fiction retraces these first "thinking machines," speculative fiction truly missed the boat on computers. The fascination with Tin Men proved irresistible; calculators did not grasp the imagination, maybe because they were too practical, too foreign, too subtly alien. When computers did make an entrance into literature it was as godlike beings, residing in vast computer-filled spaceships or covering the far side of the moon. Douglas Adams, in *The Hitchhiker's Guide to the Galaxy*, asserts that Earth itself and all life on it is a huge computer; without this knowledge, nothing here makes any sense.

Computers were large. Entire floors of office buildings were dedicated to these monsters; they needed to be super air conditioned to keep the hundreds of vacuum tubes from overheating. (A modern-day cell phone, if it used 1960s era vacuum tubes, would be as big as the Epire State Building!) Maintaining these giants required workers to wear HAZMAT-style suits, they could not afford to introduce dust to the system. It was the massive centralized computer mainframes that inspired IBM Board Chair Thomas Watson to say, "I think there is a world market for about five computers."

MACHINES GONE WILD
"WHY DO THE ROBOTS ALWAYS GET THE GIRLS?"

We use the word "robot" most correctly when we describe machines that can move on their own (usually with wheels) and can sense and respond to their surroundings, or that can respond to unique situations, like a robotic sheep shearer. Some bump into things and then randomly change course, like a robot vacuum cleaner, others transmit infrared light and respond to reflections.

Jeff of the Future Does Payroll, with a Little Help from His Friend: Piles of magnetic tapes that need to be rewound, a temperature-controlled room, a vast control panel, and a print-out for a screen. This is a past that need never be repeated.

For more mobility, spider legs are all the rage. Moving about on six or eight legs is wonderfully intricate, but trying to walk on two legs like humans involves such complexity of balance as to be impossible (so far). The rovers that landed on Mars use six wide wheels, some on arms to dislodge themselves from sticky situations.

Despite the efforts to make robots look like butlers, they, like most new technologies, were more often imagined as rapacious weapons of mass destruction. Giant steel men are just intimidating, and make for great speculative fiction. Most of the robots were actually remote control monsters; a switch in their backs or a mothership, once disabled, neutralized the robot invaders. Sort of like killing the queen bee to confuse the workers.

Guy in Robot Suit Driving Jeep: If the ride didn't loosen some bolts (above), we would have been on to something.

Remote control machines are often the precursors of robots. Remote control devices search earthquake damaged buildings, navigate volcano craters, and disarm

Computers at Your Service

MACHINES DO; MEN THINK

In 1962, futurists saw the age of computers was imminent. Children would spearhead this forward push, so the kids' book *Computers at Your Service* laid out the plan. Christmastime consumerism would be made simple with computers because: "The shopper fills out a form which asks the sex and age of the person who will receive the gift. It also asks his occupation, hobbies, and how much the shopper wants to spend. This information is punched on IBM cards and fed into the machine. The computer then prints a list of ten suitable gifts!"

To exorcise any feared ghost in the machine, kids were assured that punchcards would allow us to communicate with these metal beasts. "Luckily, we don't have to speak 'computer language.' When we feed a punched card or tape into the machine, there are tiny metal fingers that feel for the holes. Wherever there is a hole, an electrical current flows."

With a title like *Computers at Your Service*, the book affirms humans' superiority complex by boasting that, "Computers are clever because the men who made them are clever. Machines do; men think." Most importantly, computers will obey humans. "In other kinds of factories, computers have set people free from boring 'slave' jobs. These people now do more interesting, better-paying work. The machine has become the slave."

bombs. The first robot to Mars was remote controlled, but the seven-minute time lag meant the 'bot had to be programmed to make many decisions independently.

The military looks to robots for some of its missions. At this time the RC surveillance planes are used, some equipped with missiles. A far cry from the iron men with death ray eyes who haunt the magazines of yesteryear.

The smartest robots today are about as capable as ants. Mass produced, they can work together as "swarm 'bots." Some ideas: when robots touch, they exchange sections on their computer programs. Ostensibly the strongest will survive. Or the swarm 'bots simply multiply their efforts through cooperation, and even when damaged can be hauled home by members of the swarm.

Don't Touch: A self-cleaning computer (right).

Damsel in Distress!: Her chest heaves with indrawn breath (above), she screams and pounds at his merciless metal grip. Unrelenting, he blasts with his powerful gun and spears with rays from his eyes, dragging her away to his spaceship, never to be seen again.

THE RIGHTS OF ROBOTS
"'LET US MAKE THEM IN OUR IMAGE,' AND THEY SAW THAT IT WAS GOOD"

Literature waxes about robots and why they are so often shaped like humans. It's agreed that we make them in our image because we make what we know, but also because we need to feel comfortable with our creations. Jean Baudrillard wrote in his 1968 *The System of Objects*, "If it is to exert its fascination without creating insecurity, the robot must unequivocally reveal its nature as a mechanical prosthesis (its body is metallic, its gestures are discrete, jerky, and unhuman). A robot that mimicked man to the point where its gestures had a truly human fluidity would create anxiety." Just as we anthropomorphize our pets, we do the same with our machine friends.

Isaac Asimov was well known for his stance as a lover of technology. He dedicated much of his writing to the positive effects that new technologies would bring, and he asserted that robots would almost never go on killing sprees. Philip K. Dick, in *Do Androids Dream of Electric Sheep (Blade Runner)* used replicants (androids? robots? cyborgs?) to show the violence that humans are capable of. The *Star Wars* movies do the same: much of the story consists of slaughtering armies of robots. Not at all gory, is it?

In contrast, as we watch Dave silently and methodically shut down HAL of *2001: A Space Odyssey* it feels as if a ruthless murder is taking place. The catch is that HAL has been programmed to keep secret, important details of the *2001* mission; a computer asked to lie is a contradiction, one that HAL resolves by determining that he would have a better chance of completing the mission without the human crew.

Me Like Girl: The Colossus of New York (left) was a robot with a transplanted human brain, so maybe he had an excuse to want the girl.

Adam and Eve at the End of Time: These suggestively shaped robots (below) didn't seem to appreciate their booty. It was this disdain for human flesh, rather than the attraction to it, that made these metal men more menacing, more merciless, and far more alien.

Making Robots of Our Kids

JUST PRESS CTRL-ALT-DEL WHEN YOU SCREW UP

The education of the future will teach students about their history; in other words, they will study the present. What a delicious thought for the soon-to-be famous inventor! Kids will learn about new technologies, say that invisible virtual computer you're about to invent, and they will use these new inventions to study. They'll learn about you're virtual computer on a virtual computer.

The future seems to float on this spiral. New forces are domesticated (say, electricity) and comfort with these forces (wiring our houses) leads to even greater domestication, of them and of us (we get pretty used to having lights). Children surf the web, while senior citizens are baffled by the remote for the TV. The kids do animation on the computer that would have transfixed the Disney studios fifty years ago. The economy itself floats on the personal computer, just as it floats on the automobile (but not the hovercraft) and floats on the machine gun (but not the death ray) and floats on electricity (alternating current, not direct). New metaphors, new avenues opened.

Education itself went through a great democratization when it was universalized and made available for all. This in turn advances democracy, or at least advances the thought that we all can think for ourselves. On the negative side we have the warnings of Aldous Huxley's *Brave New World* and George Orwell's *1984*, where mass education becomes diluted to the needs of the rulers. Mass came to mean consumerist for Huxley and fascist for Orwell.

Today we find both these trends. In many countries it is illegal to market to children; while in the USA, Taco Bell serves school lunch. And the standardized testing that drives schools is a product of the eugenics movement, which was discredited a generation ago.

Teaching kids by rote memorization treats knowledge as something that can be poured into children. The truth is distilled and then inserted into passively accepting kids, just like programming fills a robot's computer brain. For our own comfort, we have long made robots in our image, as if they are just like little people. If education can be reduced to acquiring facts and sorting them, we may come to find that little people, our children, are just like robots.

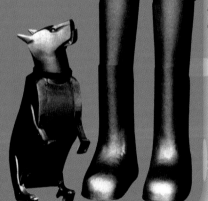

Elektro and his Pup Sparko: What an adorable pet (right), with its own pet. Elektro really doesn't walk, he rolls one foot at a time. He smokes cigarettes, mouths music, and has a hole in his heart: he's as human as the rest of us!

Robots are mostly seen as gentle butlers, and indeed that seems to be where we are headed. Robot vacuums and lawn mowers and pets (and battle 'bots) are our introduction to this new species. Smart houses might be considered home 'bots, moving food from the freezer to the oven as we enjoy our commute from work, security systems that chart our movements and light our way (and warm us and clean up after us) as we move through the house. Sounds very tame compared to sci-fi's possessed machines bent on world domination.

MACHINES AS BUTLERS
WHERE'S MY AGE OF LEISURE

The idea that machines should make our lives easier is suspect because we must now become "machine literate," able to care for and manage the ever-changing landscape of helpers.

Personal computers have allowed businesses to get rid of support staff for their executives, but often the executives end up doing the writing, layout, and printing of their own thoughts, instead of the good old days of the dictaphone. While administrative staff may be let go, a new herd of computer literates enters—programmers and systems analysts. Most ironically, it is peers who tend to teach each other computer literacy, at a huge loss of productivity and money.

Jobs themselves are changing. "The very same process of automation that causes a withdrawal of the present work force from industry causes learning itself to become the principal kind of production and consumption. Hence the folly of alarm about unemployment. Paid learning is already becoming both the dominant employment and the source of new wealth in our society. This is the new role for people in society, whereas the older mechanistic idea of

The Americans with Destructabilities Act: Perhaps the bravest thing this loner (below) could do was plead for recognition in the heartless halls of government. He could have made it through the columns, but what about those little human-sized doors?

From the Pages of *Dynamite!*: This inscrutable face (above) hides a despicable mind. Its evil plan: Serve overly sweetened beverages to dull the minds and plump the bodies of humanity's next generation.

'jobs,' or fragmer
ers,' becomes me
to Canadian cultı

And what about
"Whereas in the
had been the abs
verse is true in tl
demands the sir
discover that we
intensely involve

BIONIC
PORTAL TO (

"Cyborg" is a con
the term "cyber-
but what does it
ing "helmsman."
course) would be
as we are betwee
of years ago and
The helm is the s
modified with n
and its machine

Cyborgs could ı
pacemakers, th
expand human
cations, though
Man," who's ma
and sight are ar
hard. There are
record their eve
back that recorc
to the Internet,

Pierre Ouellette's *The Deus Machine* adds biotechnology to the horror. The AI begins as an experiment linking thousands of personal computers; the self-awareness leaps to a large computer in a biotech lab, uses the genome projects to create new life forms like huge vicious bees and toxic, fast-growing vines; it eventually gains control of weapons labs. You get the picture: world domination.

Different from AI, an IA is an "intelligent actor." This more accurately describes the auto assembler robots that can sense their surroundings and make some decisions about their actions, or the rovers on Mars that handle much of the moment-to-moment decisions but rely on Earth for the larger choices.

Intelligent actors could include the soon-to-be-released safety systems in autos, which can take command of the car if a collision is imminent. Using global positioning and relying on the swift response of mechanical sensors (milliseconds versus seconds for a poky human response), we may soon see crash-free highways. (Dream on.) The intelligence is inherent in the system, more than in a machine making intelligent choices. The car wakes up only when its "life" is in danger.

Virgin Sacrifice to the God of the Robots: The large-brained creature in the cauldron is in charge (right), but those tentacled robots, and the one with the electrode fingers, seem to be enjoying themselves too much.

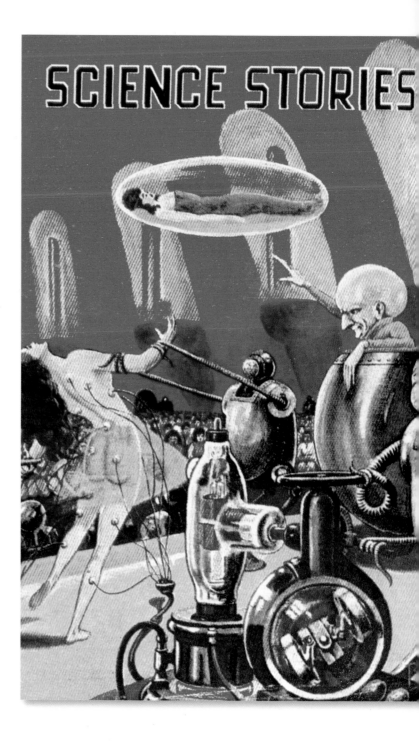

Issac Asimov's Three Laws of Robots

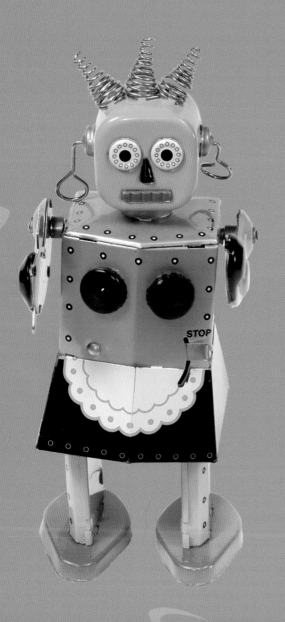

1. A robot may not injure a human being, or through inaction allow a human being to come to harm.

2. A robot must obey the orders given it by human beings except where such order would conflict with the First Law.

3. A robot must protect its own existence as long as such protection does not conflict with the First or Second Law.

"A lot of religious people think you can never exceed your maker. That's why they have such a powerful god."

—Hank Lederer, Futurist

Pierre Ouellette's *The Deus Machine* adds biotechnology to the horror. The AI begins as an experiment linking thousands of personal computers; the self-awareness leaps to a large computer in a biotech lab, uses the genome projects to create new life forms like huge vicious bees and toxic, fast-growing vines; it eventually gains control of weapons labs. You get the picture: world domination.

Different from AI, an IA is an "intelligent actor." This more accurately describes the auto assembler robots that can sense their surroundings and make some decisions about their actions, or the rovers on Mars that handle much of the moment-to-moment decisions but rely on Earth for the larger choices.

Intelligent actors could include the soon-to-be-released safety systems in autos, which can take command of the car if a collision is imminent. Using global positioning and relying on the swift response of mechanical sensors (milliseconds versus seconds for a poky human response), we may soon see crash-free highways. (Dream on.) The intelligence is inherent in the system, more than in a machine making intelligent choices. The car wakes up only when its "life" is in danger.

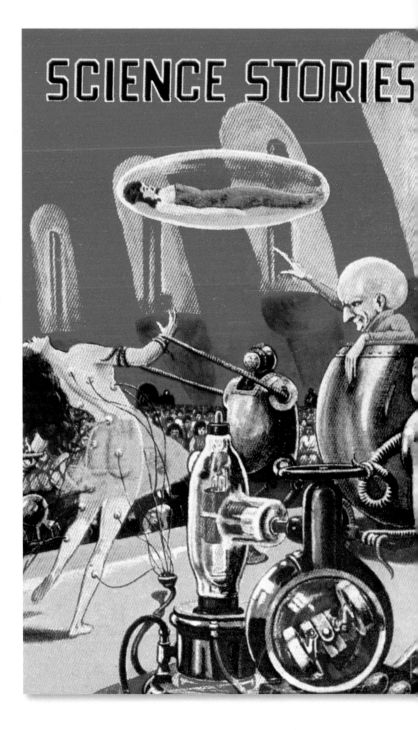

Virgin Sacrifice to the God of the Robots: The large-brained creature in the cauldron is in charge (right), but those tentacled robots, and the one with the electrode fingers, seem to be enjoying themselves too much.

Robot World

ROB-A-TRON VS. ROBOVEND

Enter "Emergency Hatch 7" and prepare for the countdown. An elevator turned "shuttle" blasts past constellations visible outside the glass hatch. After rocketing through light years of space in a minute or so, the pod hatch opens onto an alien planet, known as the second floor. Welcome to Tommy Bartlett's "Robot World and Exploratory" in the Wisconsin Dells.

Colored plastic, wires, and flashing lights are tweaked around by gangs of robots, while desperate bleeping emanates from control panels. "Caution: High Radiation Area" warns a sign overhead and another cautions visitors not to touch, with a flashing "Danger High Voltage."

Everyone out! "Radiation Alert! Security Alert! Levels are excessive. Radiation detected in an anti-matter area." A siren blares and mechanical voices strangely reminiscent of the British-accented C3PO ramble in a nervous tizzy. A robot with a Scottish accent yells, "Two minutes to melt down! Move along before the whole place goes up like a Roman candle!"

Obviously these robots' inventor, Tommy Bartlett, envisioned that robots would be doing more than the dishes, cleaning our toilets, and just being our friends. These good robots would prevent the world—and our intergalactic spaceships—from radiation overload. Except, of course, if they met the evil robots bent on domination of the universe.

Some robots, like Zord Robovend, are to be trusted. "Hi kids! My name is Zord, I want to be your friend. I have a special surprise I picked just for you!" Robots never lie, right? Another devilish robot challenges weak earthlings to a battle to the death. "I am Rob-A-Tron from the planet Zircon. My order is to seek out inferior humans. You have three chances to disarm me...good luck, humans!"

After this dizzying encounter with Tommy Bartlett's 1970s-era vision of our future with robots, it's with odd prescience the sign at the exit declares: "The body may become weak from lack of use."

Tank da Vinci: Shedding projectiles as it rolled (above), or just spinning to provide fresh armor to face the enemy, this fanciful tank looked mysteriously like a flying saucer and sported a lookout turret with excellent shielding. And look how fast it could have moved!

"People react to fear, not love. They don't teach that in Sunday school."

—Richard Milhous Nixon (1913–1994), U. S. President

Inventing Away War
GYROSCOPIC TANKS WITH RADIUM RAYS

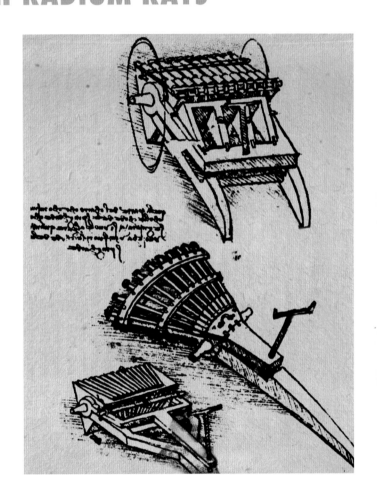

CAN MACHINES DO OUR DIRTIEST WORK FOR US?
ROBOT VS. ROBOT

What a sweet dream indeed that we should invent our way out of war. Maybe this is the greatest scientific folly of all.

The advent of electricity came at one of those rare moments in European history, a period when spheres of influence had been established and various governments found peace to be in their mutual self-interest. The U. S. was not yet a global presence and its Civil War had slaked its passions for war. No wonder that, when colonial wars cropped up at the end of the 1800s, the popular imagination looked to the newly discovered forces and the new machines they powered to end the human tragedy of war altogether.

It was an age of horses and rifles and bayonets, when deaths on the battlefield were caused as much by trauma as by infected wounds and diarrhea. Relegating the violence to machines seemed a wonderful solution.

Machines making war on each other would allow humans to continue their peaceful lives. It would let civilization

Rapid-Fire Ballista: da Vinci well understood the power of miniaturization in this early machine gun (above). In this case, many small projectiles work far more efficiently than a monster cannonball. Note the triangular cannon supports, allowing for at least three volleys before reloading.

progress, exorcised of this demon. Economies could flourish without the devastation that war regularly imposes. All that was needed was a fleet of machines to decide the bureaucratic squabbles between nations. Of course an open battlefield would be needed, an arena where these "gladiatrons" could blast away at each other.

At the time, the colonies seemed vast beyond anyone's needs and could serve this purpose well. For example, the British proved their superiority as they mowed down spear-carrying Zulus in South Africa. Now that civilization had been established, couldn't ugly lopsided battles like this be settled without the carnage?

During the decades of peace, war was a "game of statecraft." Whichever nation was best at diplomacy and negotiation won in this new civilized world. But the utopian vision didn't last.

Maybe it is folly to believe that war can be invented away. Maybe it was psychology, the new science of thought, that most subtle of energies, which was first to pose the question: Is war innate in humanity?

It was economics, another science of subtle forces, that soon asked whether war was actually a necessity for a healthy economy, a prerequisite of progress. Has there ever been a civilization free of the stink of war? Is war an evil mutation of development, or is it the compost in which the most advanced societies thrive?

Fantasy War: Ah, the orderliness of modern warfare! Those fanciful wings could never stabilize against the recoil of a machine gun (left); imagine how the balloons would flip once their undercarriage cannons were discharged!

MODERN MACHINES ATTACK THE MASSES
REINVENTING TOTAL WAR

Some of the first modern weapons were imagined by the consummate inventor Leonardo da Vinci. Shrapnel was named after an officer in the British army, but da Vinci's designs of pineapple-shaped explosives leave no doubt that he saw the value of flying bits of hot metal. He designed a ballista that fired multiple bolts—a precursor for the machine gun. To protect the good guys, da Vinci developed mobile armor that is as comical a tank as we are likely to see.

Warfare took a decidedly grim turn in the First World War. Not only were machine guns and new technologies like poison gases used to clear the stalemate of trench warfare, but eventually tanks and bomber airplanes and radar entered the fray (gee, maybe war does spur development). Most importantly, the concept of total war made a re-entrance into the European theater.

Napoleon understood total war. In the early 1800s the French used a highly decentralized army, living off the land as much as possible, to inflict deep casualties on the economy of the enemy. War in the "colonies" (i. e., nine tenths of the planet) often was a gruesome display of firepower arrayed against poorly equipped warriors, with the civilians, often referred to as savages or natives, usually taking the brunt of the battle. Just as in ages past, siege warfare was devastating to everyone stuck behind the walls. Looking back even further, many armies practiced total war; the battles in the Bible are harrowing in their bloodletting.

But the concept of "innocent bystander" was maybe a new one, from the quaint notion that when armies fight one another they should focus on killing soldiers, not on burning and pillaging. Maybe the innocent bystander was an idea based on the new political reality—the nation-state and its citizens—that superceded the king and his lieges. Maybe this new idea was based on the economics of business as separate from government.

It was the innocent bystander who believed that war was simply a game, who surmised it could be played in an arena, and who dreamed that it could be mechanized to the point that humans and their edifices could be spared. The neutron bomb that killed all the people and left the buildings intact was only somewhat successful. Afterwards, the victors still had to clean up all the radiation and corpses before colonizing the abandoned cities.

The return of total war by Napoleon Bonaparte rained fire on the cities of humanity. Ludwig van Beethoven was so impressed by Napoleon's futuristic promise to rid Europe of the tyranny of old monarchies that he wrote his third symphony to honor the French leader. Upon seeing the carnage of the power-hungry dictator, Beethoven renamed the symphony "Eroica," or heroic, "to celebrate the memory of a great man" who was corrupted by power.

The Russians knew they were no match for Napoleon's war machines, so they simply abandoned Moscow and armed released convicts with lots of gunpowder and fuses and told them to burn the occupied city to the ground. Napoleon's modern cannons couldn't fight this guerrilla warfare and eventually left the city defeated. Modern armaments couldn't defeat clandestine terrorism by determined anarchists.

To finally eradicate violence on humans but recognize our innate need to go fight, why not just have all these modern war machines fight each other? The obvious next step in sci-fi was that the machines would become so powerful they would dominate humanity. There was no arena for war anymore. The machines would make war on their masters.

GIGANTO-GYROSCOPIC
DEATH STARS ARE BIG TARGETS

Gigantism is what happens when things are well. When bugs ruled Earth they measured ten to twelve feet long and weighed hundreds of pounds; now much of the world dines on their tiny relatives. The dinosaurs grew to gigantic proportions and did quite well for millions of years. Now their offspring chirp to us as they flit from branch to branch. Power expresses itself in big ways, before it subtlety exerts itself. Gigantism is eventually followed by its opposite: miniaturization.

Sci-fi scribes have dreamed of city-sized war machines to roll, float, or fly through space and wreak havoc on peaceful Luddite worlds. Generals and admirals have asked for more cash to build bigger war machines, like frustrated three year olds setting up every tin soldier they own, then rolling over them with their trikes. One such design, the submarine-land *Dreadnaught* tank, zooms out of the water and appears ready to roll over anything but the cliffs of Dover.

During the "War to End All Wars," the humble bicycle was a very reliable way to move troops about. Take the subtlety of gyroscopic balance, weaponize it, make it gigantic, and you rule the battlefield. The *Electro-Gyro Cruiser* envisioned during WWI was one such giant. This twenty-story-tall motorcycle tank balanced thanks to the hidden powers of gyroscopes. Cannons and machine guns (they hadn't yet conceived of the flame thrower) were fitted on every opening to blast anything that moved on the ground.

Another deadly beast based on gyroscopes was the *Gyro-Electric Destroyer*, which added spikes to its unicycle wheel. Not only did this allow for better traction, but brave enemies could be squashed and impaled as the

A War-in-One: Gigantism at its best, the name said it all. The *Submarine-Land Dreadnaught* (below) appeared to stretch from Dunkirk to Dover and crush everything in its path.

Spinning Blades of Death: Okay, so it couldn't really fly (far right), but get a load of the debris trail.

THE NEW
Science and Invention
IN PICTURES

THE SUBMARINE-LAND DREADNAUGHT
See Page 966

25 Cen

IN PRIZE $12,000 FOR PICTURE
See Page 968

Destroy! Destroy!: A few details needed to be worked out for the *Gyro-Electric Destroyer* (below), such as how to absorb the recoil of the cannons, and how to climb into the cockpit. But imagine if several of these met up on a battlefield: "Neerowm! Kabloosh. Dadadadadada…"

Gyro or "Yee-ro": Does your suburban assault vehicle have this kind of clearance? Spitting hot lead as it burned rubber (bottom center), this was one fantasy war toy. Note how it combined the amazing (as yet untested) invisible powers of electricity and gyroscope.

wheel turned. Imagine the dust that would fly, however, if some sneaky Prussian put a stick in its spokes!

Generals in the early 1900s regarded trench warfare as the pinnacle of conflict and often lacked the imagination that created much more effective warfare that stealth bombing and guerrilla warfare would later inflict on humans. The ultimate weapon that remained on the drawing board, therefore, was the trench destroyer. This gigantic beast balanced between two wheels, and ironically easily became mired in the trenches they were destined to clear.

HIDDEN FORCES
ELECTRIC BLASTS AND RADIUM RAYS

Miniaturization took off as inventors tried to digest the powers behind electricity, radium, and the radio. Radar was invented in the latter half of WWI; using invisible radiowaves, radar can see through clouds and through the dark to reveal the baddies and expose them to our merciful guns. During WWII, Britain was apparently able to counter the German's radar, diverting their bombers miles from the intended targets. Britain then used German spies they'd detained to report home that the bombing raids had been devastatingly accurate.

The powers of electricity gave way in the 1920s to the imagined powers of radium. Electricity expresses itself as zigzagging through space, as unpredictable in its aim as lightning, though it can be imagined that the *Tesla Destroyer* would work invisibly to detonate distant battleships.

Scientists envisioned that radioactivity would be beamed as straight as light rays. We find the gigantic radium destroyer, followed by the miniaturized ray guns that could take out airplanes, or airplanes that could split tanks (the submarine-land *Dreadnaught* for example) with death rays.

Death Rays and Jet Packs: These paratrooper warriors (above) of the future have it all. Watch out, zeppelins.

Death Rays: A blast from an electric beam sends sparks along the wings of a warplane (right). This was almost probable, and shows a level of understanding of the disruption such an attack would have on the delicate innards of a plane. A few years later, *Amazing Stories* showed a massively overdone death ray shearing tanks in two. Do you wonder how deep a gash that would leave in the earth?

From the invisible to the highly visible: The explosions of atomic warheads in two cities in Japan brought a new face to war. The era of total war had reached a crescendo, and the individualists among us decided they'd survive such a blast in their own underground bunkers. The bomb shelter became the badge of honor for the husband who truly cared about his family. "Duck and cover" was not enough for these do-it-yourselfers.

Just as the flamethrower is a crude and horrific form of a death ray, so is the modern use of radioactive elements a crude use of this subtle force. Uranium 235 is a waste product of nuclear reactors, but is also an amazingly dense metal. It is not used as a nuclear weapon but rather as an explosive cannonball that is so heavy it penetrates armor, with much of the radioactive metal vaporizing upon impact.

Electro-Aero Destroyer: In this 1939 photo (top left), even a beautiful model could operate the new "anti-aircraft" target gun. Energy rays zapped out of the barrel in a "pulsating electric beam," making metal bullets and gunpowder obsolete.

Light Amplification by Stimulated Emission of Radiation: Only a dollar! And its cousin the phaser, also a dollar! Nuclear radiation experiments were two dollars, while harmless science projects were only 35 cents (left). The humble laser, a step down from the death ray, found its greatest application in the laser-pointer. Oh, and in reading compact discs.

The Tri-Beam Cosmic Ray Gun: Finally junior could play Buck Rogers in the backyard and foil the dastardly schemes of the neighbor kids with his handy death ray (bottom left).

Heed the Voice from Above: An army's psychological operations rely on spreading the news to both friend and foe (top far right). These Royal Air Force blimps from 1939 dangled loudspeakers to broadcast all the news that was approved for release. Maybe some Led Zep anthems, too.

ZEPPELINS TO ATOMICS
SHOOT THAT BIG TARGET IN THE SKY

The airplane is one of the greatest successes of science and engineering. Nothing else (well, maybe the microwave oven) captures the imagination like the power of flight. Flight has a couple of drawbacks, namely taking off and landing, and these remain fruitful fields for exploration.

The zeppelin had not quite been put to rest in 1902 when the British War Office commissioned the *Aeroplane Airship*. Six propeller engines, each mounted on a wide wing, propelled the blimp at a velocity of twenty miles per hour. The wings provided some lift, the hydrogen-filled balloon the rest.

The combination of a helicopter and an airplane soon gave way to rocket-assisted launches. Then came the ejection seat, a rocket-assisted launch for the hapless pilot from a debilitated plane.

The Bomb Shelter: The indomitable American spirit prevails (below). Setting aside their baseball mitts, this team began excavation of their future home, pausing only to show off for their barber/local civil defense commander. No girls allowed.

The Fallout Shelter: In the event of nuclear winter, be prepared with canned spinach and board games (below). Bring a small shovel just in case. No whining allowed.

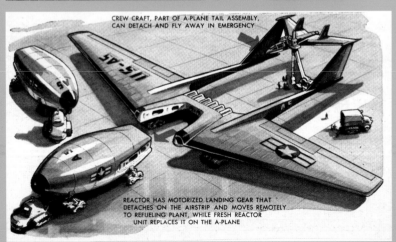

CREW CRAFT, PART OF A-PLANE TAIL ASSEMBLY, CAN DETACH AND FLY AWAY IN EMERGENCY

REACTOR HAS MOTORIZED LANDING GEAR THAT DETACHES ON THE AIRSTRIP AND MOVES REMOTELY TO REFUELING PLANT, WHILE FRESH REACTOR UNIT REPLACES IT ON THE A-PLANE

Up, Up, and Away: Retractable rotary systems (top left) meant airplanes could lift off like helicopters, then fly like planes.

Flying Stove Top: The precursor of the flying saucer, this wingless wonder (center) looked great on paper but....

The Flying Nuclear Reactor: Keeping the skies safe was never so simple (bottom left). The reactor could be exchanged and even dumped in mid-flight.

Rocket Assisted Lift-Off: Using a rocket to get a jet into the air (far top right) resulted in three things: a great appreciation for effects of acceleration on the human body; the invention of the seat belt; and the creation of Murphy's Law.

Flying Atomic Aircraft Carrier: Gigantism wins in this seaplane (far center right), which carried its own escort of fighters and bombers as it patrolled the globe.

Atomic Airplane-Boat: "This flying boat (far bottom right) could 'bury' radiation hazard in water." Assuming it was flying over water.

At the same time, wing designs changed. The bat-winged airplane seems a cousin to the modern stealth bomber. The folding wing, while neat in pictures and while storing in hangars, was not so smooth in flight.

Ah, the atomic plane, nothing like a flying nuclear reactor to bring peace of mind to the people. The A-plane conceived by Northrop Aircraft would have a detachable nuclear reactor at the airplane's nose; the crew would pilot the plane from an escape craft positioned between the twin tails. The long-nosed version was the most common design, but the scale of the plane would make it a giant flying aircraft carrier. Smaller planes could land on its back and elevators could draw them into storage. The aero-carrier could ferry smaller planes across the globe and remain aloft as a fueling station. What a great target for a heat-seeking missile!

PSYOPS AND THE MONSTERS OF MODERN WAR
WINNING HEARTS AND MINDS THROUGH ESP

The police of the future were envisioned as lone, armored units delivering stun rays. There were rumors at the 1968 Democratic Convention that the police used sub-sonic noises to terrify demonstrators, inducing not only panic but also widespread bowel movements. Tear gas, water hoses, and rubber bullets have created a more humane form of crowd control. Together with pepper sprays and electric tasers, however, they raise new questions about cruel and unusual punishment.

The "Army of One" is a strategy that matured with Napoleon. He sent his troops into the countryside to pillage and live off the land, with orders to gather a few days hence at some important site. Much imagination has gone into creating the ultimate fighting machine, the robocop, the invincible soldier equipped with the most amazing collection of firepower, endurance, and shielding. Modern armies have reduced the time it takes their soldiers to get medical care from days to hours. Along with the development of armed robots there are great strides in robotic arms as prosthetics. Imagine what could be done if sci-fi dreams of artificial blood were a reality!

The trauma of war to soldiers has lessened greatly in the past centuries, though training someone to kill without compunction always has a cost. The trauma of total war on the "innocent" is yet to be determined.

Modern warfare includes a great amount of psychological warfare, designed to inflict debilitating terror in enemy civilians and soldiers. Resistance is futile to "shock and awe." It's not hard to imagine that for the designers of high-tech military hardware—from microrobotics to death rays—the ultimate fear is another invisible force: the sorcery of psychic weaponry. American Cold War red baiters worried about the Ruskies brainwashing us and beating us in the ESP battle, as demonstrated in the book *Psychic Discoveries Behind the Iron Curtain* by Sheila Ostrander and Lynn Schroeder. What kind of brain shield can protect us from telepathy?

"Crime will be virtually abolished by transferring to the preventative process of schools and education the problems of conduct which police, courts, and prisons now seek to remedy when it is too late."

—National Education Association, quoted in "What Shall We Be Like in 1950," *The Literary Digest*, January 10, 1950

Our Very Own Saucers: Humanity's indomitable spirit will prevail (far top left). We will someday have our own flying saucers, most probably flown by the military.

The Unimaginable: For all our scientific measurements and technological control (far bottom left), there are still forces that defy our understanding.

The Brain—Our Second Most Important Organ: Floating suggestively in outer space (right), this great brain houses our mutual psychic powers. Humankind rises to the challenge, and faces down our invisible enemies with latent psychic powers.

A PAPERBACK LIBRARY SCIENCE-FICTION NOVEL

62-554 50¢

BEYOND THE SPECTRUM

The cosmic war of the centuries where Earth's psychic powers battle the invisible invaders from the planet Nihil

Martin Thomas

Unleashing the Beast Within: Floating in the depths of space, our intrepid hero and her toothy sidekick (above) appear to have beckoned a beast to protect them. Is this our mind power or our lust on the loose?

Relativity is the Propaganda of Counter-Revolutionary Ideology

SOVIET SCIENTISTS WEIGH IN

$E = MC^2$ did not fit the Soviet party line. The *Astronomical Journal* of the Soviet Union in 1940 dismissed Albert Einstein's genius as anti-Marxist garbage. "The theory of a relativistic universe is the hostile work of the agents of fascism. It is the revolting propaganda of a moribund, counter-revolutionary ideology."

After the bomb was dropped, the Cold War was heating up. Superior Soviet scientists rejected theories of their counterparts on the other side of the Iron Curtain. Russian Professor M. S. Eigenson wrote a 1950 article in the U. S. S. R.'s *Nature Priroda*, called "The Crisis of Bourgeois Cosmology," which stated that: "Einstein's finite cosmos, closing back in on itself, is a product of capitalistic idealism intimately related to the decline of bourgeois culture and a return to the outmoded cosmic models of Ptolemy and Aristotle."

Marxist government was thought to lead to utopian bliss, and capitalism a throwback to the savageries of the Stone Age (or the Greeks, at least). Marx's studies of history focused on materialism, the economic building blocks of how we get commodities like food, houses, and tools; of primary interest was who controls the means of production, and with this in mind a vast retelling of history takes place. Einstein's relativity hocus-pocus was seen as not grounded in concrete materialism by the Soviet scientists.

In 1920, western physicist Niels Bohr wrote the "Principle of Complementarity," unveiling the ultimate nature of the electron as a mystery described either as a wave or a particle. When combined with the "Principle of Uncertainty," (we can never know the exact position of an electron because our measuring influences the electron) a great deal of metaphysics is introduced into physics. These apparent contradictions did not sit well with the Soviet scientists of the time. In 1949, physicist Maurice Comfort mocked this typical capitalist inconsistency, writing that: "The task of dialectics is not to accept the contradictory proposal that an electron is both a wave and a particle.... So far as bourgeois physical theory is concerned...the atomic nucleus constitutes, as it were, the central knot of contradiction of the physical world, just as the simple commodity does in the economic world...[and] bourgeois theory in physics is no more capable of understanding the nature of the atomic nucleus than bourgeois theory in economics was capable of understanding the nature of commodities."

Scientist Andrei Zhdanov went a step further and theorized that Bohr's theory on electrons was western propaganda, using atomic structure to support immature bourgeois ideas of liberty. In *Literaturnaya Gazeta,* Zhdanov wrote in 1948: "The subterfuges of contemporary bourgeois atomic physicists lead them to conclusions about 'freedom of the will' of electrons." Another gentle reminder that our passions for the truth need to be understood before the truth will be revealed, and that the truth itself will be transformed over time as its contradictions overtake it.

This advertisement (below) idealistically illustrates the idea that a pound of fuel will light Chicago's vast skyline.

Gulliver Can't Believe His Eyes: A flying city of mad scientists (left)! This is too good to believe.

Sunday Drive: Feel the loft, feel the speed (far right). Love the bare-chested guy and the magic dust in the pod's wake. Monorails and pointy skyscrapers from the city on a pedestal pierce the sun in this 1950s sci-fi pulp.

Cities of the Future
HOMES IN DOMES

THE DOMESTIC MARKET
THE HOME STAGE IS SET

It wasn't a lord with servants who first thought of making a washing machine or a dishwasher. It was either a working woman or the beleaguered husband of a stressed homemaker who saw the need for labor-saving devices.

Once the floodgates opened, thousands of inventions sought patents—the electric iron and its sidekick the fold-down ironing board, the clothes wringer, clothes spinner, clothes washer, clothes dryer, the pressure cooker, the slow cooker, the microwave oven.

The vacuum cleaner had a poor entrance into the marketing field. In those days advertisements often took the form of demonstrations on stage, before a vaudeville act or a debate (family fun!). The curtains lifted on the first vacuum salesman. He dumped a ceremonial amount of dust and lint on the floor, and with great fanfare turned on his "vacuum." With a great blast it cleaned the floor, blowing the dirt off the stage and into the air. The floor was clean, the air filled with grit, and the audience was coughing as they fled the theater. Vacuum sales that day:

zero. Another century passed before this original home cleaner had a rebirth as the leaf blower.

With the medical discovery of germs and the realization that filth can lead to things like cholera epidemics, cleanliness became an obsession. The indoor flushable toilet was one such invention, using a continuous flow of water to cleanse the toilet bowl. Public officials worried whether the water supply could handle such a drain, however, until the harmonious sounding "syphonic flush" was invented by the Brit Thomas Crapper (or his employee, Albert Giblin).

The chemical toilet was created in the 1930s, used today mostly by airlines and construction sites the world over. The bidet is a further advancement of personal cleanliness, using water to clean the nether regions, after wiping with paper.

Toilets are also the source of the first modern pictograms. The pants-wearing male, the dress-wearing female lavatory symbol prepared the popular imagination for the cross-walk symbol. It was the Olympics in the '70s that refined these international symbols to include skiing, bobsled, skating, and so on.

THE CULT OF CLEANLINESS AND ITS DIRTY SECRETS

BYE-BYE BUGS

The future is sanitary. No longer will bugs and bacteria muddle up our lives, in fact, sci-fi and futurists rarely mention these microscopic scourges.

Our obsession with cleanliness—from the continuous flush toilet to antibiotic hand soaps—has its costs. Those first toilets were clean but wasteful; antibiotic soap cleans the hands only slightly better than a good scrubbing, yet enables bacteria to develop resistance to its effects.

"As early as the 1920s and 1930s, epidemiologists were finding that the crusade for cleanliness at the beginning of the century, far from combating polio, was promoting it.... Epidemics became most severe where standards of plumbing and cleanliness were highest. There, young people were first exposed to the virus long after the end of maternal immunity,"

Ultrasonic Ultra-Squeaky Clean: Developed by Sanyo in the late 1960s (left), this ultrasonic bath was deemed a, "human washing machine," and surely wouldn't fit in any ordinary bathroom.

The Cuteophone: More important than seeing your granddaughter on the phone is showing off your new outfit (right). The technology for the videophone has been available since the 1970s, but who wants to show off bed-head or bad outfits at all hours?

(**1**) Gutters and leaders—everlasting — they're lead. (**2**) Cames of lead give these windows real beauty. (**3**) Clapboards are protected with white-lead paint. (**4**) Lead casing gives complete protection to the underground electric service wires. (**5**) The interior walls are beautified with sanitary white-lead paint. (**6**) Woodwork, too, is preserved and beautified with pure white-lead paint. (**7**) Water supply line fittings are made tight with red-lead. (**8**) Even the lining of the picturesque flower box can withstand endless years of weathering. It, too, is made of lead.

Four walls—a roof—and LEAD

CAN lead be turned into gold? Look at this picture. In it is the answer to the old alchemist's dream. For today man has done more than transform dull pigs of lead into so many glittering nuggets. In his effort to beautify and protect the four walls and roof of his home, he has discovered that lead is the more useful, and therefore gladly exchanges the gold for lead.

The hand-wrought gutters and leaders, for example, are not only beautiful, but weather can't wear them. They are made of lead. The old craftsmanship of the rustic casement windows will remain unchanged through the centuries. All the cames that hold the glass in place consist of lead.

Another form of lead, one in most general use today, is present in this home. It doesn't look like lead—yet it is made from lead and has the metal's superior qualities of endurance, weather resistance and protection. You'll find it on the clapboards and trim—on the interior walls and woodwork. It is the basic carbonate of the metal, called white-lead, which makes a paint that gives both beauty and protection to the surface.

There are many other unseen uses of lead in this home. Lead helps to give the glass of the electric light bulbs their transparency, also the fine glass tableware its brilliancy. Lead is in the glaze of the chinaware and in that of the bathtub and sink. And a lead device makes it safe to telephone when lightnings play.

National Lead Company makes lead products for practically every purpose for which lead is used today. If you would like to know more about this wonder metal of many uses, just write to our nearest branch.

NATIONAL LEAD COMPANY

New York, 111 Broadway; Boston, 131 State Street; Buffalo, 116 Oak Street; Chicago, 900 West 18th Street; Cincinnati, 659 Freeman Avenue; Cleveland, 820 West Superior Avenue; St Louis, 722 Chestnut Street; San Francisco, 485 California Street; Pittsburgh, National Lead & Oil Co. of Pa., 316 Fourth Avenue; Philadelphia, John T. Lewis & Bros. Co., 437 Chestnut Street.

The V. P.—Is "the president" too threatening?: This 8-track of dictaphones (above) invoked the inventing power of Edison and allowed plenty of time to puff a cig while chatting to the secretary—via magnetic tape.

according to *Why Things Bite Back: Technology and the Revenge of the Unintended Consequences* by Edward Tenner.

DDT is a wonder drug, and it was sprayed liberally over cities and armies alike in WWII to combat pests and the diseases they carried. DDT was far cleaner than the arsenic, lead, and copper poisons it replaced. But by 1947, there were flies resistant to DDT, and by the mid 1950s, "…body lice in many parts of the world were already unaffected by DDT treatment." What a folly that our pal "*Popular Science* foresaw 'total victory on the insect front'" soon after the war.

A more modern saint in the cult of cleanliness is wall-to-wall carpeting. Our modern buildings are often too airtight, and adding flooring made of plastic fibers and glue may not be healthy. The EPA itself removed 27,000 square feet of carpeting in 1978 from its headquarters and cured its "sick building syndrome." Beyond the vapors off-gassed from carpet, there are the allergens from dust mites that may be causing asthma, one of our most common chronic diseases. And our friend the vacuum cleaner, unless refitted with excellent filters every couple of months, is, just like its leaf-blower cousins, simply lifting the filth from the floors and suspending it in the air. So much for a germ-free future!

LABOR SAVING OR LABOR ELIMINATING
ROBO-SECRETARIES

"A secretary that never talks back," was the selling point of the robot answering machine. Yes, machines will rule the world (along with the techies who maintain them).

The dictaphone lets you get so much more done. From the briefcase-sized dictation-transcription machine (typing on paper with its erase-o-matic belt) to the advanced dictabelt desktop dictaphones to the "most carryable" V. P. voice writer, these were essential for the aspiring vice president.

Okay, so the secretary wasn't discarded, "she" was promoted to administrative assistant whose first task was to transcribe the musings from the boss's dictaphone. Carbon paper and typewriters—shudder—were vast improvements over quill and ink, but it wasn't any specialized tape that rid the world of typos. It was the computer.

The frustration with machines has always been with us. A precursor of the computer was the calculator, a glorified cash register with a printer added. To trust our financial calculating to machines took a leap of faith that was aided by the accountants who took to these counting machines, secure in the knowledge that their ability to interpret, crunch, and spin the numbers would never be mechanized.

So why aren't we all talking on videophones? Interesting question, one most of us can answer if we put our minds to it. The technology has been available for decades, but perhaps no one wants to show their face on a bad hair day. Imagine if your boss could check in every time you call in "sick." Today, the main use of video messages is talking to family. Kids talking to grandparents is the market here, which is not the most lucrative demographic. So there is little reason for videophones, other than the need to view the cuteness of the person you're talking to.

AGAIN *plastics* TAKE A FRONT SEAT!

Brain Plug: Work while you sleep (far left)? Just as swindlers convinced consumers that French could be absorbed through headphones while catching some Zs, so too have scientists dreamed of plugging in the brain like a computer and downloading all the world's knowledge.

Big Money in Plastics: Laugh off the muddy dog prints in your new car (above). All thanks to plastic.

Steel House: "Instead of an annual paint job, you'll be giving your enamel house a wash job (left)."

THE HOME OF THE FUTURE
IF YOU BUILD IT...

The Victorian wooden houses were deemed firetraps after too many cities had burned to the ground (darn cows kicking over lanterns). New materials were needed for the future city.

Thomas Edison made a very down-to-earth choice when he marketed concrete houses. Like many of the modern wonder materials, concrete is very pliable and can be formed in almost any shape. "I feel so *secure* because it's concrete, *fire-safe concrete*." Many Edison concrete houses, while never the success he'd imagined, still stand today.

"Buddy, I've got one word for you: plastics." Young Benjamin Braddock of *The Graduate* never took seriously this advice from his mentor, but plastics have endured as one of the most "plastic" of building materials. While synonymous with tacky, the power of plastic is yet to be realized. How many of you are in buildings with polyvinyl siding? If you have a child, chances are you're surrounded with plastic toys and your child is sitting in a plastic diaper.

Lead has been admired for centuries. The ancient Romans intoxicated themselves with lead beverages; the pewter mugs of America's founding fathers were an alloy of tin and lead. "Four walls—a roof—and LEAD," boasts ads from the 1950s. Yes, lead is a wonderful material. Pliable and powerful, it was the plastic of an earlier generation. Lead paint is durable and beautiful, with a smooth finish that modern paints rarely match. Yet the health effects of lead ingestion are devastating and widespread. Imagine our children growing up free of lead poisoning, possibly the first generation to do so in many years. Think you can outsmart grandpa, eh?

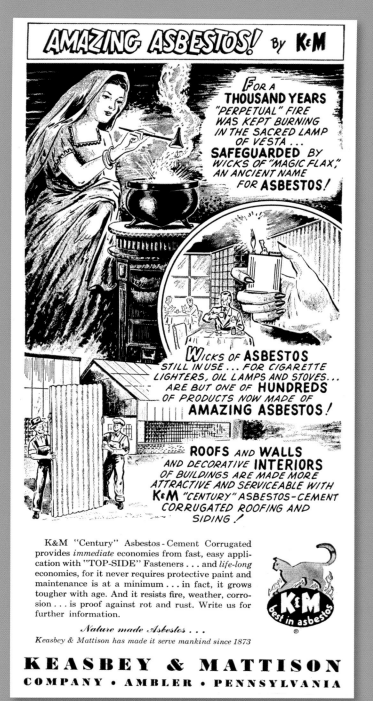

Penetrating Paradise: Plato's Atlantis was always thought to be a product of his imagination, but U. S. Representative Ignatius Donnelly thought otherwise. Ever since Donnelly proposed that this utopian world existed in his *Atlantis: The Antediluvian World*, sci-fi has melded the classical world with the futuristic colonies (top).

Modern Living in Foam!: *Design News* (bottom), "The House would meet the most exotic tastes of even the Mongol Emperor known for his love of splendor."

Asbestos is a naturally occurring mineral that is pliable and fibrous enough to be woven into fabrics. It is amazing for its ability, like most minerals, to handle high heat without bursting into flames. But the fibers themselves are what make asbestos deadly: they become airborne, are inhaled, and sadly persist in the lungs, tearing at the delicate structure within for decades.

Asbestos' fire-retarding properties made it perfect for use in housing: roof shingles, siding, insulation, wallboard. It was used in car engines and made excellent gaskets and brake pads. It was just so easy to use! "This amazing material can be sprayed on any ceiling, no matter how intricate the design or how irregular the surface." In fact, it is still used in many of the ways that plastics are, for making water cisterns and tubs, for roofing and siding. And it's natural!

Asbestos is toxic, and so is lead; plastic isn't natural. Why not build your house of steel? Better yet, after rolling out the steel, coat it at high temperatures with porcelain enamel. The walls (and ceilings) are thin but extremely strong. Maybe some asbestos for sound proofing?

The 1960s saw the birth of a house made entirely out of foam. Xanadu, the "Foam House of Tomorrow," was built in Florida and the Wisconsin Dells as a tourist site to show visitors how they'll be living in, say, 2006. Finally, architecture had discarded the stifling right angles and everything followed curves that would have made Spanish architect Antonio Gaudí purr. Hardened foam offered instant insulation but the problem remained that it was just, well, gross. Xanadu in the Wisconsin Dells amused decades of tourists until the property was sold to put in another water park. The Foam House of Tomorrow was thrown away in the municipal dump and, thanks to the miracle material, will be there maybe forever.

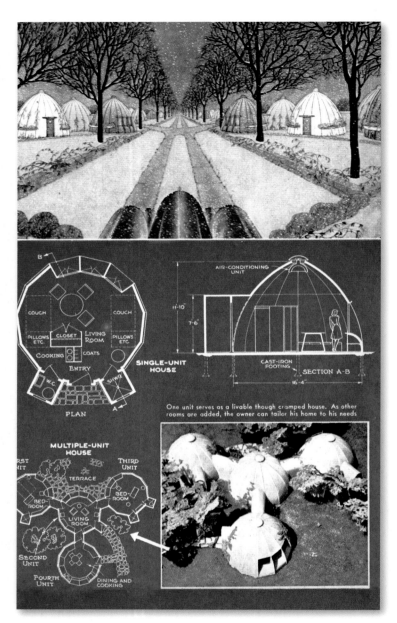

PLAN

SINGLE-UNIT HOUSE

AIR-CONDITIONING UNIT

SECTION A-B

COUCH COUCH

PILLOWS ETC. CLOSET LIVING ROOM PILLOWS ETC.

COOKING COATS

ENTRY

W.C. SHWR

CAST-IRON FOOTING

One unit serves as a livable though cramped house. As other rooms are added, the owner can tailor his home to his needs

MULTIPLE-UNIT HOUSE

FIRST UNIT THIRD UNIT

TERRACE

BED-ROOM BED-ROOM

LIVING ROOM

SECOND UNIT

FOURTH UNIT DINING AND COOKING

Metal Pods: Would you live in a stainless steel igloo (above)? For only $670 would you crash here? Keep your stuff in the pod, spend your life in the yard.

HOMES IN DOMES
INSTANT PAPER HOUSES

"You can't make an electric light by perfecting a wax candle," said Harvard professor Max Wagner in *Popular Science*. Why make a steel house square? A starter home could be a single room with added cubicles for showers and a toilet. As the family grows, add more pods for a full kitchen, bathroom, and bedrooms. "Later on, if the family decreased in size, unneeded rooms could be sold."

The buckydome was the house of the future thirty years ago. Buckminster Fuller, poet and sage, will ever be remembered for the domes he designed. His ideas went far beyond airplane hangars and sports facilities.

Fuller imagined a dome house made of high-quality paper. The sections, with instructions printed right on them, would be stapled together in an hour or two. Add a "black box" that would double as toilet, composter, and fuel source, and you've got a moveable hearth.

DOMED OR DOOMED CITIES
WEATHER ALWAYS PERMITTING

The myth of Atlantis has a staying power that is awe inspiring. Underwater cities have long been part of popular imagination, whether as dwellings for fairies and mermen, or as hopeful oases for loved ones lost at sea.

Returning to the sea is alluring, an imagined return to safety and to the calm beneath the storm. The walled cities of Europe and the walled manors of the Mideast were

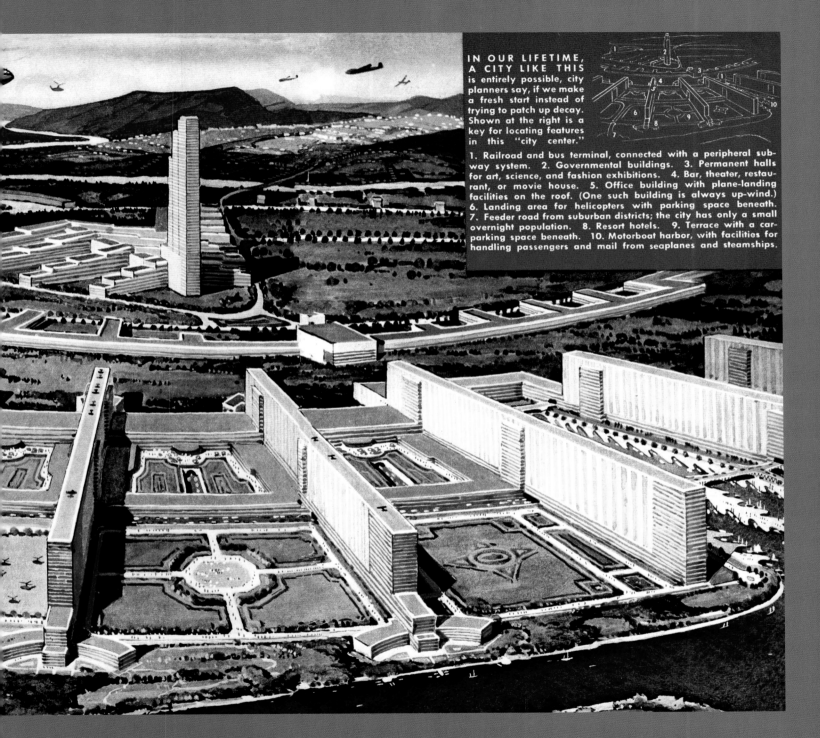

IN OUR LIFETIME, A CITY LIKE THIS is entirely possible, city planners say, if we make a fresh start instead of trying to patch up decay. Shown at the right is a key for locating features in this "city center."

1. Railroad and bus terminal, connected with a peripheral subway system. 2. Governmental buildings. 3. Permanent halls for art, science, and fashion exhibitions. 4. Bar, theater, restaurant, or movie house. 5. Office building with plane-landing facilities on the roof. (One such building is always up-wind.) 6. Landing area for helicopters with parking space beneath. 7. Feeder road from suburban districts; the city has only a small overnight population. 8. Resort hotels. 9. Terrace with a car-parking space beneath. 10. Motorboat harbor, with facilities for handling passengers and mail from seaplanes and steamships.

PLANNED UTOPIAN COMMUNITIES

THE FAMILY IS OBSOLETE

A "planned community" today brings to mind a gated suburb of townhomes set within a perfectly sterile landscape of lollipop trees. Urban planning was actually a populist movement to get people out of the deadly soup of city living. The automobile was truly a tool of independence for many. As the cities became more crowded, horse traffic multiplied to the point where in New York City more than 300 million pounds of manure were produced annually, spreading tuberculosis and tetanus. Traffic jams inspired horses to panic, kick, and bite; according to *Why Things Bite Back* by Edward Tenner, "Horses died after only a few years of service, usually in the middle of the street, up to 15,000 a year in New York."

Inspired by Frank Lloyd Wright's designs from the 1930s, utopian towns put "carports" on the side of the house rather than the old-style stable/garage in the back alley. The car was to be shown off, and soon giant two- and three-car garages were accessible by a giant driveway filling the front yard. The Victorian front porch entrance was obsolete and motorists went in the side door.

Levittown, New Jersey, was the first of these planned communities. Abraham Levitt made thousands of prefab houses, mostly for vets returning from WWII combat.

City planning was not a new idea, but the concept of creating an egalitarian society and living situation based on high-minded morals flourished in the 1800s. During the Industrial Revolution in Britain, Welshman Robert Owen purchased the Chorlton Twist Company near Manchester.

To prevent Dickensian horrors of the downtrodden workers, Owen set up his factory to treat all the workers fairly, educate their children, and create a part agricultural/part industrial community. He took his collectivist ideas to America and set up a successful socialist colony, attracting members to his earthly utopia with a social order based on equality of work and profit and community ownership. He purchased the Indiana town of Harmonie from German religious utopians in 1827; they had owned property communally and vowed a celibate life, which didn't bode well for their birth rate. Owen's dream colony was a communal success—at least until the other community members objected to his antireligious views.

The U. S. is full of early experiments in communal living. The self-sufficient Amana Colonies in Iowa preached that personal possessions were *verboten*, all profits were shared, goods were available to everyone, and groups of forty people would share meals rather than individual families. The Shakers, who were named because they would shake with religious fervor in church, advocated celibacy, sharing all material goods, but anyone could opt out of their group. Brook Farm in Massachusetts was another famous socialist experiment set up in 1841, made up of transcendentalists who agreed that everyone should receive the same wages, have the same housing and food.

The Oneida Community in upstate New York advocated "Perfectionism" to abolish sin and make its members perfect. To achieve this enlightened state, the private family was abolished and members lived communally in complex marriages in which every man was married to every woman. In the best eugenics tradition, the community decided which couples would produce the best offspring and let them breed accordingly.

The 1960s saw a revival of communes by hippies across the country. Meanwhile in the East, Soviet state farms and Mao Zedong's collective farms gave the idea of utopian communes a black eye. Pol Pot and his Khmer Rouge forced their dream society on Cambodians by outlawing religion, abolishing money, banning private property, restricting marriage, and forcing city dwellers to farm in order to create a classless society.

On the other hand, multinational corporations such as Wal-Mart have followed the idea of the working planned community used by Henry Ford in Detroit and transferred the idea to China and other cheap labor markets. While these factory communities don't abolish marriage, by any means, the workers sleep in communal dormitory bunkbeds and are discouraged from leaving the barbed-wire compounds. Robert Owen and other utopian idealists would not be amused.

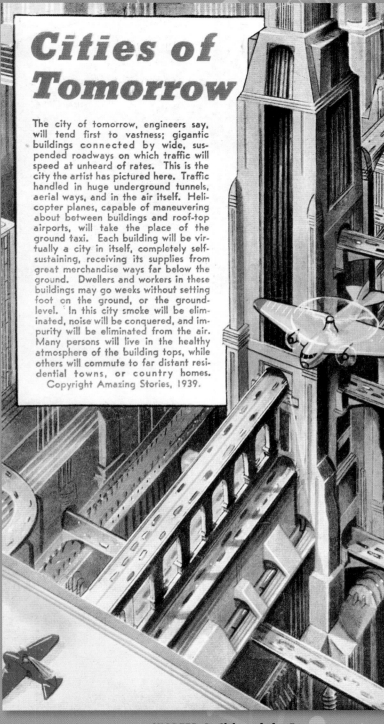

Cities of Tomorrow

The city of tomorrow, engineers say, will tend first to vastness; gigantic buildings connected by wide, suspended roadways on which traffic will speed at unheard of rates. This is the city the artist has pictured here. Traffic handled in huge underground tunnels, aerial ways, and in the air itself. Helicopter planes, capable of maneuvering about between buildings and roof-top airports, will take the place of the ground taxi. Each building will be virtually a city in itself, completely self-sustaining, receiving its supplies from great merchandise ways far below the ground. Dwellers and workers in these buildings may go weeks without setting foot on the ground, or the ground-level. In this city smoke will be eliminated, noise will be conquered, and impurity will be eliminated from the air. Many persons will live in the healthy atmosphere of the building tops, while others will commute to far distant residential towns, or country homes. Copyright Amazing Stories, 1939.

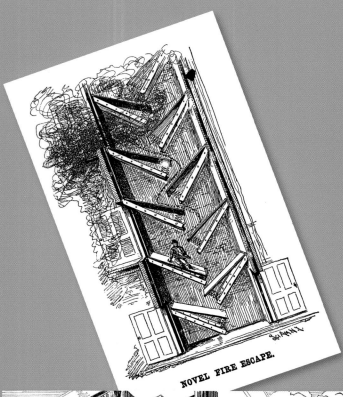

NOVEL FIRE ESCAPE.

Fire!: At least someone is being practical (top left). In the city of the future—with all those enormous buildings, buzzing with aircraft—what about safety? Perhaps inspired by dangerous carnival rides, this drawing from a 1903 *Scientific American* shows how the bowler-hatted gentleman is bounced to the bottom.

Planes Above, Gardens Below: In between office cubicles (bottom left), autos race through tunnels carved through skyscrapers. Crash-free highways would need to be invented first.

Dad, You Look Stressed: It can't be that hard to drive these floating pods (bottom right) or they'd never be allowed. Note the thousand-foot fall if he swerves too far left, that is if the elevated roads actually do anything. Why build roads if car pods float?

THIA OF THE DRYLANDS by Hal Vincent

Other Scientific Fiction by:

nothing compared to the protection afforded by the seas. The author Jules Verne described submarines and worlds within our world; modern Japanese author Kobo Abe describes a return to the sea as the oceans rise and the world outside becomes inhospitable. The return to the sea, the desire to tame the abyss, is also a dream of reclaiming wholeness, of integrating our liquid emotions with our physical bodies.

Doming a city is another dream. Milan domed an intersection of streets with a lovely "Galleria," and a century later, the modern covered shopping mall was invented. The Mall of America is a small commercial city indoors, so why not just cover a real city and keep in the perfect weather? Like an underwater bubble, like an egg, the dome rises delicately but inevitably out of the landscape. Often imagined as transparent, it is the ultimate Garden of Eden: beautiful, clean, and openly visible to all, this paradise is shielded and nurtures the human spirit.

UTOPIAN HIGH-RISES
EVERYONE TO NIAGARA FALLS

The utopias of yesterday are often fascinating in their desire for control and uniformity. The Panopticon was a nineteenth-century design for a prison where all inmates were watched from a central location; these pre-designed cities flushed away the stink of human activity by planning everything perfectly, a fresh start based on that most exact of sciences, urban planning. "There will be only one tall building in the hub.... It will house the government for Democracity," bragged *Popular Science* about the "World of Tomorrow" exhibit at the 1939 World's Fair. Such trust in the government and in the founders (the guys who built these models) in such a contained world. Visitors to this model city watched from a revolving platform, a god-like

City in a Bottle: If cities are going to float above Earth (top), why not just send them out into space? With light years to the next solar systems, these city-planets could fly through the Milky Way and generations hence could visit other planets. One problem: how would these domed cities produce gravity to keep their citizens grounded?

Elevator to a New World: No cars in this world (bottom), where men are men and women are quite feminine. Note the glow holes in the dome and the fields in the distance.

Controlling the Weather

SHELTER FROM THE STORM

Only mopey dystopians thought we'd have anything but perfect weather in the future. Along with a cure for the common cold, scientists would finally put the reins on Mother Nature's penchant for hurricanes, tsunamis, lightning, and other foul weather that would ruin the picnic of the future. Some futurists assumed that we'd just have to dome our cities to keep out the storms and regulate sunshine, but in that case, why not just put these new metropoli underwater?

The first sign of weather control began with cloud seeding, dropping silver iodide or dry ice from planes into clouds to lower the temperature and produce precipitation. There is evidence that airplane jet contrails do influence the weather, but the effectiveness of cloud seeding has been questioned and some states have outlawed rainmaking because it could cause shortages elsewhere.

Before airplanes were used to drop the chemical concoctions into the clouds, Charles Hatfield gained notoriety around the country as the "Rainmaker." Born in Minnesota in 1875, Hatfield soon realized that his talent was more sought after in the dry climates of the Southwest.

When Los Angeles was in the midst of a drought in 1905, Hatfield was hired for $1,000 to bring the rain. He stirred up a batch of mysterious chemicals and let them evaporate through an enormous tower up into the sky, and the rains came. The angelic city's thirst was quenched as water reservoirs of L. A. rose eighteen inches.

Newspapers reported that Hatfield was paid to perform his chemical rain dance 500 times for drought-stricken areas. His most notorious miracle, however, was when he was hired by the city of San Diego for $10,000. Once again he raised his tower and sent his chemicals into the cloudless sky. Nothing happened and the city officials smelled a rat. Hatfield and his brother quickly mixed up another brew of their secret chemical recipe and finally the rains came. In fact for two weeks it stormed. Rivers spilled their banks, streets flooded, dams crumbled under the weight of the water. Fifteen inches of rain fell and several people were killed by the disaster. Although Hatfield the Rainmaker became a household name from the event, he was stiffed by the city of San Diego, which was sued for millions of dollars in damages by its own citizens.

Europe had its own plan for controlling the clouds, "weather shooting." The enormous cone-shaped *Kanitz Weather Gun* was loaded with gunpowder and simply blasted at a cloud to prevent damaging hail and produce rain instead. A 1901 article in *Scientific American* boasted that 20,000 of these weather guns had been stationed in France, Germany, Hungary, Switzerland, and Italy. The breech-loading mortars were about thirty feet long with a charge that, "...is a metallic cartridge of blasting powder. After the discharge a loud, shrill whistling is heard, lasting about fourteen or fifteen seconds.... The Italian government has such faith in weather-shooting that it supplies wine-growers with powder at the rate of three cents a pound."

gaze that exposed all to its cleansing gaze. Eerily similar to the Panopticon may be the subtext that we'd be also gazing in interest at each other, exposing the deviants among us.

So essential to the dynamic of an economy, as we now understand it, is the ability to innovate, to respond to needs, for each generation to make its dreams reality. In other words: To make a mess.

Frank Lloyd Wright weighed in with his dream of the ultimate high-rise for Chicago. In 1956, Wright unveiled his plan for a mile-high building with a twenty-six-foot-tall model. Towering four times the height of the present day Sears Tower, Wright's behemoth would be serviced by atomic-powered elevators that would shoot residents to the top. Fifteen thousand parking spots would fill the base of the building, while 150 helicopter landing pads would jut out from the top. All puny buildings on the ground around this structure would be leveled for parks. He demonstrated to skeptics how the building wouldn't sway in spite of Chicago's famous winds, and he conveniently avoided mentioning any fire hazard. As part of his "Mile-High Illinois" plan, ten of these skyscrapers would be built and supply as much office space as all of New York City. With a $100 million tag at the time, Wright hesitantly admitted that, "No one could afford to build it now, but in the future no one can afford not to build it."

If building walls is addition, then arches are the architect's algebra, and domes are advanced calculus. The organizing of forces to create such order is one testament to the human spirit, and it was celebrated in the enormous glass-walled domes of the World's Fairs. In about 1900, King Champ Gillette (yes, that's his name, the guy who marketed the disposable safety razor) imagined using the power of Niagara Falls to create enough electricity to power his dream city.

Gillette's *Metropolis*: "Under a perfect economical system of production and distribution and a system combining the greatest elements of progress, there can be only one city on a continent, and possibly only one in the world." —Gillette, King Champ, *The Human Drift*

GETTING AROUND TOWN
WHERE ARE THE ZEPPELIN HIGHWAYS?

Hidden forces surround us. We know they exist because we surf them, whether they be our own heartbeats, the sleep-wake cycle, the flow of our thoughts, or the feel of sunlight and the force of the storms it creates.

The World's Fairs lifted towers dozens of stories tall. The futurists of the day imagined much greater constructions, skyscrapers dwarfed by their offspring, the massive almost-cities that approached from just beyond the horizon. How would these edifices hold themselves up? Leave that to the natural momentum of engineering and technology.

How will puny humanity fit into this vastness? We'll walk, of course. Skyways and moving sidewalks link the upper levels of the city, maybe after the fear of heights is overcome. On ten-lane-wide thoroughfares autos run cleanly and safely below, within and around the glowing buildings. And the ultimate in multimodal transit, airplanes and zeppelins touch down far above. All these fantasies are linked among the hanging gardens. Myths will come to life.

When the Eiffel Tower was built in Paris, it was to be the first dirigible airport. Cities across the world lined up to have the next skyscraper airport. The Empire State Building was built with a landing platform (now the observation deck) made as an unloading spot for zeppelin passengers. In an effort to turn Minneapolis and St. Paul into a new destination, city planners announced "An Eiffel Tower in Minneapolis" in 1889. Critics scoffed at this Minnesotan "Tower of Babel" that was proposed to be 2,000 to 3,000 feet in height. "…elevators and drive-ways, enabling the visitors to make use of horse carriages inspecting it, as well as restaurants and amusement establishments along the road, are included in the elaborate plans…and thus our city would gain world fame."

In spite of many of the old dreams of our cities' futures, much of this has come true, except for the walking, the cleanliness, and the zeppelins. Can we dream our way out of it? Forget the science, let us believe we'll find hidden energies that will give us floating pods, floating humans, and floating cities.

Something Eerie: The city becomes oppressive and impersonal in this image (top far left) from *Metropolis.*

Condo for Rent—Hi-Class, Lo-Grav Neighborhood: With the artificial gravity of this space station's spin (far bottom left), a very comfortable living area is created. The higher up the sides of this torus, the gravity becomes slightly less. The view might be better, too.

Moon Base: Agriculture and energy production are crucial to any extraterrestrial home (center). Well, the moon has no water and the cost of transporting the massive amounts needed to start some sort of hydrological cycle is enormous. Mars may be the first extraterrestrial agriculture for this reason.

A·MENTAL·ASSASSINATION·REVEALED · AS EXEMPLIFIED WITH THE CO-OPERATION OF THEIR·IMPERIAL·MAJESTIES THE EMPEROR AND EMPRESS OF ALL THE RUSSIAS IN THE IMPERIAL PALACE GATCHINA, Nov. 17TH 29TH 1884.

From Sketch by Matt Morgan

A. S. SEER'S UNION SQUARE LITHO. PRINT, N.Y.

MR· WASHINGTON·IRVING·BISHOP The First and World Eminent · THOUGHT·READER·

Medical Marvels
FOOD PILLS AND CLONE FARMS

THE END OF PAIN
JUST HOOK ON THESE ELECTRIC WIRES

In no other field have so many promises been made to an eager populace as in the field of medical marvels. Food will come as tablets and offer complete nutrition. Obesity will be eradicated by fat-eating pills. ESP, telekinesis, and psychic surgery will replace the necessity of cutting open patients in the surgery room. Clone farms with your own personal DNA embedded will produce replacement organs when your ticker goes amok. Then when (or if) you actually do keel over, just quick freeze your body and you can be thawed in 100 years, reanimated (thanks to the miracle of cryogenics), and your old age ailments will be cured with medicine from the future.

Before the twentieth century, good medicine hurt. Patients were bled, purged, and fed awful remedies. George Washington was treated for "quinsy" by doctors who insisted he drink many vials of mercury. The macho cry, "Pain is weakness leaving the body," could have been the motto of the day.

As the mystery of the body slowly revealed itself with the dawn of the modern age, scientists realized how human-

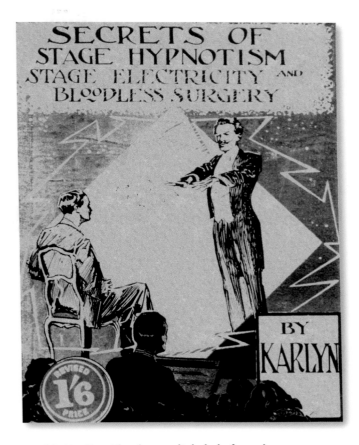

Psychic Healing: Thanks to a little help from the man behind the curtain (far left), psychics could determine patients' prognosis and even prevent a stab in the back.

Bloodless Surgery: Seeing is believing, and besides how would anyone know back in 1912 if the surgery worked or not (above)? The future of medicine promised no more cutting up of the body, but rather a return to "hands-on" healing.

ity could also be manipulated. Control of our species was ripped from any higher power and put into our own hands. Eugenicists from Nietzsche to the Nazis proposed careful breeding of the population to ensure a better tomorrow. In 1922, Dr. Albert Edward Wiggam bragged in *The New Dialogue of Science* that, "Had Jesus been among us, he would have been president of the First Eugenics Congress."

Anything seemed possible and any cure was in reach. Magnetism, vibrations, radiowaves, radioactivity, and electricity were unseen forces that were harnessed to cure what ails you. Thankfully, the Federal Drug Administration began regulating drugs in 1906 and medical devices in 1938. Still, the public was open to the pitch of inventive snake oil salesmen and prepared to pay for their dangerous gadgets. Most of all, patients hoped for a healthy future free of pain.

THE TRAVELLING QUACK.

Snake Oil Salesman: This 1889 caricature (center) of a medical quackery showed that the public was already skeptical of miracle cures in little vials. When "modern cures," such as magnetism, electricity, and radioactivity, were deemed to be new cure-alls, the sick would try anything for relief.

A Tall Man is a Successful Man: In the era of advertisements boasting how tall men were the only lucky ones with the ladies (right), this gadget sold in the early 1900s supposedly added to a man's height.

Atomic Farming

GLOWING PLANTS, DEAD WEEDS

With miracle chemicals killing all the bugs and weeds around the crops, the motto of "better living through chemistry" spread the belief that simply fiddling with molecular structure or spraying DDT would do the job.

Pesticides and herbicides were soon rivaled by radioactive isotopes to aid farmers of the future. The 1960 children's book *The Walt Disney Story of Our Friend the Atom* by Heinz Haber detailed how radioactive cobalt could raise healthier—or at least larger—crops: "Active atoms are a great help to scientists looking for better ways to grow our food. In careful tests, they let the rays from active cobalt shine on growing corn. They found that the rays changed the seeds from the corn. Plants that were grown from some of those seeds were better than any that farmers ever had before.... Tests like these show how to get bigger and better crops."

"I doubt whether Darwin will have the determination to survive a difficult journey of several years. My studies of physiognomy indicate that people with a broad, squat nose like his do not have the character."

—Robert Fitzroy (1805–1865), Captain of the *Beagle* and governor of New Zealand

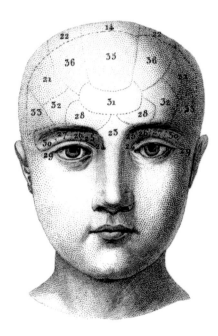

Read Your Skull!: While your palm revealed your destiny (above), your cranium revealed your personality. Phrenology had disciples across the sciences until the brain was slowly understood. The brain may be compartmentalized, but bumps and valleys on the skull didn't necessarily denote brilliance or idiocy. *Collection of Museum of Questionable Medical Devices*

PHRENOLOGY AND PHYSIOGNOMY
BUMPS ON THE HEAD MAKE YOU SMART

The early map of the brain was a study of the shape of the skull, called phrenology. Bumps and depressions on the noggin could reveal the character of the person beneath the cranium.

In 1796, the original phrenologist Dr. Franz Joseph Gall lectured that the, "…brain is an organ of the mind." He claimed to have distinguished twenty-seven faculties of the brain, including destructiveness, amativeness, philo-progenitiveness (desire to have offspring), combativeness, and secretiveness, all of which could be discerned by mapping a person's skull.

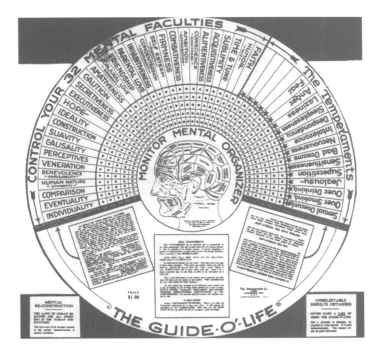

There are two powerful truths in these ideas. First is the most simple, that the brain is an organ of the mind. In the 1700s, the seat of the soul, and thus of consciousness, was the heart. Phrenology put thoughts into the head. Note further that Dr. Gall did not limit the mind to *only* the brain. Science now teaches that many hormones are neurotransmitters. The chemistry of the mind—neurotransmitters, endorphins, and the like—resides throughout the nervous system *and* in the endocrine system of glands in several places outside the brain. The mind may be bigger than just the brain.

The second truth is that different portions of the brain are dedicated to different functions. Dr. Gall talked of character housed in different areas of the brain, while we speak of sight and hearing and higher thought centers. We owe this breakthrough to the "science" of phrenology.

Ironically Dr. Gall was found, upon his death, to have a small tumor in his cerebellum, in the site where he would have diagnosed amativeness. His cranium was found to be twice the normal thickness, so he was especially thick skulled.

Huge phrenology machines, with a frightening metallic cap, measured the skull and gave "scientific" readouts of the new and expanded list of thirty-two personality traits and mental faculties—including sexamity, suavity, and sublimity. Phrenology machines deduced that Caucasians were the superior race, so who could argue with this modern scientific marvel? Yet even to this day the study of physiognomy (the shape of the face, the nose, how close set are the eyes) is marketed in popular literature as a traditional Chinese method of diagnosis.

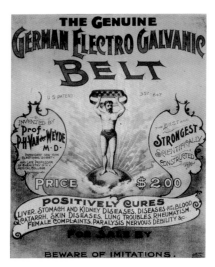

Shocking Clothes: The constant buzz of electricity (left) was thought to cure just about anything. Electric hairbrushes were promoted to spur a healthy coiffure. The most popular of all the quackery devices were the electrical belts to jolt you to life. Most models had battery acid packs, but some featured a plug-in to the wall.

VITA RADIUM SUPPOSITORIES

Actual Size of Suppository

Stay Regular with Radium!: Efforts to get "curative" radioactivity in the body involved drinking, inhaling, injecting, rubbing on the skin, and finally inserting in the rectum (above). The result: "…on the human body is like recharging has on an electric battery." As a further caveat, the ad warned, "And remember, radium taken into the system remains for months, continuing its curative restorative work. Thus, the effects are NOT merely temporary."

RADIOACTIVE RX
TAKE TWO RADIUM SUPPOSITORIES AND CALL ME IN THE MORNING

When Marie Curie discovered radium around the turn of the century, everyone wanted a bite. Her husband, Pierre Curie, was the first guinea pig to volunteer and suffered severe burns after binding his arm with radium salts. He contracted radiation poisoning, an unknown condition at the time. Once the Curies teamed up with doctors to show that cancerous cells were more vulnerable to radioactivity, radium was thought to be the new cure-all.

At Battle Creek Sanitarium, cereal and medical magnate John Harvey Kellogg set up radium inhaling stations to get the radioactivity as deep into the body as possible. Radium water dispensers swept the country—with manufacturers from New York to Los Angeles—to cure everything from asthma to apoplexy.

Ointments and lotions were laced with radioactive elements to seep into the skin if breathing and drinking it didn't do the trick. The "Magik Radium Massage: when massaged into the sex parts acts as a healthy tonic and stimulant, especially to the testes…the parts begin to take on a more healthy glow and appearance, and shrunken tissues begin to fill out and become plump." The treatment with the deepest visceral impact, however, was the 1926 radioactive Vita Radium Suppositories to help, "…restore failing manly vigor and overcome apathy in women."

"X rays will prove to be a hoax."

—Lord Kelvin (1824–1907),
British scientist

MARITAL AIDS OF THE FUTURE
VIBRATORS, DILATORS, AND ENLARGERS

Before Woody Allen's *Orgasmatron* and sci-fi's robo sex slaves, scientists and quacks turned the latest technology into bedroom helpers. Victorian prudence was passé as doctors prescribed electrical belts, mechanical pumps, and radioactive ointments to stimulate the nether regions. After all, everybody wants the future to ooze eroticism and sex.

Doctors discovered the "abdominal brain" that had been ignored by squeamish puritans. Thanks to the miraculous *G-H-R Electric Thermitis Dilator* foot-long rectal probe and prostate gland warmer from 1918, your buns could be warmed to a toasty 100 degrees. Combined with a special ultraviolet comb complete with anal and penile accessories, your sex life would never be the same.

A giant strap-on phallus "in a class by itself" was the intimidating *Recto Rotor,* which spun and shot lubricant from its vent holes. Ad copy boasted that it offered: "The Quick Relief of Piles, Constipation, and Prostrate Troubles. The RECTO ROTOR is the only device that reaches the Vital Spot effectively.... Large Enough to be Efficient, Small Enough for Anyone Over 15 Years Old." Prudishly it advised: "It is used by the patient himself in the privacy of his own home."

Other modern sex aids, thanks to new miracle materials, offered a sort of penile sling for rigidity in the days before Viagra®. The *Wimpus*, the *Erector*, the *Robut-Man*, and the *Monster Auto-Man* let a man perform his manly duties "Where there is no erection at all."

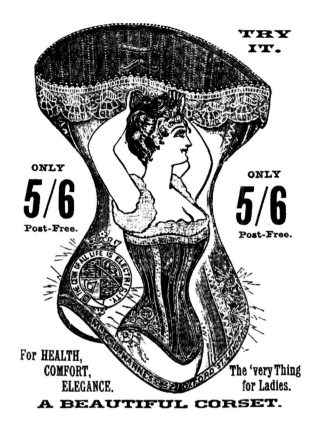

TRY IT.

ONLY **5/6** Post-Free.

ONLY **5/6** Post-Free.

For HEALTH, COMFORT, ELEGANCE.

The 'very Thing for Ladies.

A BEAUTIFUL CORSET.

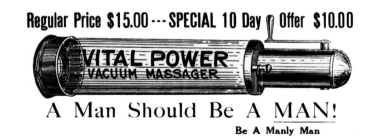

Regular Price $15.00 --- SPECIAL 10 Day Offer $10.00

VITAL POWER VACUUM MASSAGER

A Man Should Be A MAN!
Be A Manly Man

Be a Manly Man: Get pumped and ready for action! Harping on every man's fear of not performing his manly deed to his utmost, this ad (above) for the *Vital Power Vacuum Massager* held the secret to becoming bigger than life. *Collection of Museum of Questionable Medical Devices*

Men embarrassed by that sinking feeling could also pursue the "New Road to 'Strongville': The Greatest Boon to Life" by encircling their privates with Dr. Kratzenstein's *Potentor* ring to "Get Back Your Pep, Vim, Vigor, and Vitality." More common, however, were elaborate hand pumps that encouraged customers to "Be a Manly Man" and crank up "The Vital Power Vacuum Massager [which] Invigorates, Enlarges Shrunken and Undeveloped Organs. It is impossible for a woman to love a man who is sexually weak. To enjoy life and be loved by women you must be a man. A man who is sexually weak is unfit to marry. Weak men hate themselves. Upon the strength of the sexual organ depends sexual strength, in both men and women, furnishing the ambition and energy for all advancement in life. It is a well-established scientific fact that musicians, financiers, and pugilists are men of exceptionally strong sexual power."

On the other hand, women who were uninterested in sex were simply considered "hysterical." The answer around the turn of the century was weekly visits to the clinic for special check-ups in which handy doctors relieved "hysterical paroxysm" through genital massage. When a British doctor invented the vibrator in the 1880s, in one wave of his wand he saved physicians hours of tedious treatment. Soon women of the future didn't need a doctor to cure their hysteria as home remedies were available in the form of giant plug-in vibrators looking like mechanical drills, and do-it-yourself vibrating chairs.

Cure for Pain: Electricity and pain do not mix, according to the wisdom of this ad (right). By buzzing your body with a small electrical current, your symptoms were relieved. Either that or your body became numb. *Collection of Museum of Questionable Medical Devices*

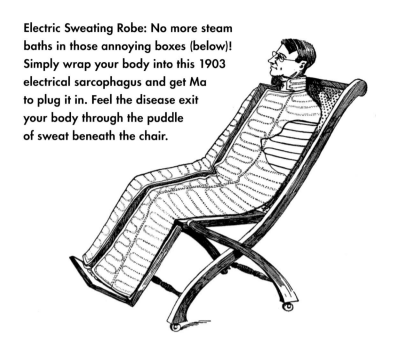

Electric Sweating Robe: No more steam baths in those annoying boxes (below)! Simply wrap your body into this 1903 electrical sarcophagus and get Ma to plug it in. Feel the disease exit your body through the puddle of sweat beneath the chair.

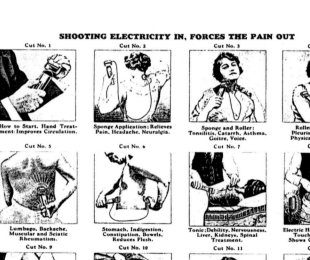

SHOOTING ELECTRICITY IN, FORCES THE PAIN OUT

Cut No. 1 — How to Start. Hand Treatment: Improves Circulation.

Cut No. 2 — Sponge Application: Relieves Pain, Headache, Neuralgia.

Cut No. 3 — Sponge and Roller: Tonsilitis, Catarrh, Asthma, Goitre, Voice.

Cut No. 4 — Roller Application: Pleurisy, Weak Heart, Physical Development.

Cut No. 5 — Lumbago, Backache, Muscular and Sciatic Rheumatism.

Cut No. 6 — Stomach, Indigestion, Constipation, Bowels, Reduces Flesh.

Cut No. 7 — Tonic: Debility, Nervousness, Liver, Kidneys, Spinal Treatment.

Cut No. 8 — Electric Hand Massage: The Touch that Soothes. Shows Complete Outfit.

Cut No. 9 — Concentrated Application: Rheumatism, Stiff Joints, Soreness.

Cut No. 10 — Electric Bath: Skin, Health, Poor Circulation, Tired Feet.

Cut No. 11 — Scalp: Dandruff, Falling Hair, Brain Stimulation, Restfulness.

Cut No. 12 — Hair: Fluffy and Lustrous, Promotes Luxuriant Growth.

MYSTERY FORCES IN THE BODY
TOUCH YOUR TOES TO CURE YOUR CANCER

Push the appropriate spot on the bottom of your feet and ta dah! Your sore throat is cured, your cancer is receding. Reflexology probably began in the early 1900s with Zone Therapy created by Dr. William Fitzgerald and Dr. Edwin Bowers. They divided the body into ten vertical zones that had a sort of energy wire jutting straight into the skull. Their 1917 book started a whole new fad of pushing various parts of your body to cure the corresponding ailment.

Chinese acupressure and Japanese shiatsu massage date back hundreds of years and describe in elaborate detail the flow of *ch'i*, or energy, through the body. They are more focused on improving wellness than curing disease, yet more than fifty illnesses can supposedly be cured, including high blood pressure, diabetes, prostate problems, hemorrhoids, and so on. The only problem with acupressure—as with its big brother acupuncture—is that results vary, and patients seem to be prone to the power of suggestion, or placebo effect. Acupuncture is reportedly used as an effective anesthetic for quite invasive surgeries. More to the point, it is accepted in many hospitals and covered by most health insurance plans.

Acupressure underwent a revival in the 1980s when applying pressure to the correct zone of your wrist was deemed a cure for seasickness. Sea Bands were marketed by a British company as a wristband to cure motion sickness. Today reflexology has updated its terminology to "meridians" rather than antiquated "zones" and the hand and ear have been found to have as many pressure points as the bottom of the feet. So now we know why Pontius Pilate incessantly washed his hands.

"Until there is a practical alternative to blind trust in the doctor, the truth about the doctor is so terrible that we dare not face it."

—George Bernard Shaw (1856–1950), Irish playwright

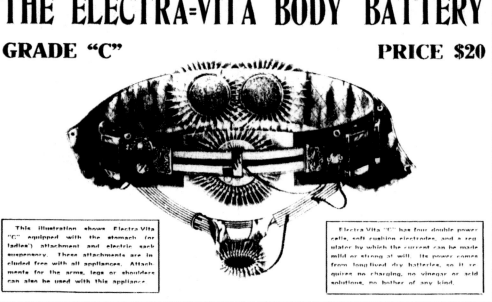

THE ELECTRA=VITA BODY BATTERY

GRADE "C"   PRICE $20

This illustration shows Electra Vita "C" equipped with the stomach (or ladies') attachment and electric sack suspensory. These attachments are included free with all appliances. Attachments for the arms, legs or shoulders can also be used with this appliance.

Electra Vita "C" has four double power cells, soft cushion electrodes, and a regulator by which the current can be made mild or strong at will. Its power comes from long lived dry batteries, so it requires no charging, no vinegar or acid solutions, no bother of any kind.

Sex, Youth, the Future

IF YOU LIVE LONG ENOUGH, YOU'LL LIVE FOREVER

Sex and youth are integral parts of visions of the future. So much of the artwork features beautifully endowed young women and ruggedly youthful men.

The future promises eternal youth (finally the Conquistadors will find their fountain). What an irony the idea of being forever young holds, because it's a fact that for all of us the future is a time when we'll be older than we are now. Yet we so often see the future as freedom from aging, freedom from aches and blemishes, where every injury and every disease is vanquished.

Indeed that is still the goal. We imagine a time when we push back the average lifespan by more than a year every year. In other words, if you're fifty-two, your projected lifespan might by eighty-two; by the time you're fifty-three, you're slated to live until eighty-three and a half. Nothing can stop us if we just live long enough to enjoy the future.

A Dog's Life: The ultimate pets (right), these little guys are as loveable as their mistress. Actually, with that outfit, that necklace, and hair clasp (and those curves), she may be simply a robot of another sort.

DRINK YOUR PEE
BODILY FLUIDS DO A BODY GOOD

Just as Jesus Christ put his saliva on a blind man's eyes to restore his sight, so can you too benefit from the power of spit. At least that's what some quacks have proposed. Modern Indian pop guru Deepak Chopra recommends scraping your tongue then spitting in a glass of water. Take that mixture and rub it in your eyes to cure cataracts.

Fellow countryman Morarji Desai, India's former prime minister, bragged to Dan Rather in 1978 about regularly drinking his own pee. He gave urine drinking credit for living past ninety-nine years old, and claimed that Indian ayurvedic medicine was overlooked by Western countries who were obsessed with modern cures. Ayurvedic medicine is now a big health food business in the U. S., but such cures as drinking donkey urine to cure epilepsy haven't caught on.

The idea of ingesting bodily fluids, or "drinking your own output," is not just an Indian tradition, however. Pliny the Elder wrote that the most curative liquid was urine from a virgin boy, and added that ingesting goat dung in urine is a remedy for anorexia, alcoholism, and, of course, nausea. In a rebuttal to modern medicine's obsession with chemical compounds and scientific observation, naturalists proposed that our bodies already know how to cure themselves but we are simply wasting these golden showers. Even the founder of modern dentistry, Pierre Fauchard of Paris, recommended rinsing the mouth with urine to stop toothache pain.

> "The medical virtues of man's urine, both inwardly given, and outwardly applied, would require rather a whole book, than a part of an essay, to enumerate and insist on."
>
> —Robert Boyle (1627–1691), British scientist

THE MOST HIGHLY PRIZED
Feminine Attractions

Are: A Beautiful, Rich Complexion; Bright, Sparkling Eyes; A Sweet, Pure Breath; A Graceful, Well-Developed Figure; Vigorous Mental Action; and a Vivacious Manner.

These attractions are all characteristic of healthy young womanhood—natural to her normal estate. If a young lady does not possess them, it is safe to say, with absolute certainty, that she has some disordered condition of health—that she is trespassing upon, or neglecting, some hygienic law, whose strict observance is necessary to her maintenance in health and beauty. In the upper ranks of society, and especially of fashionable society, the fairer sex are particularly prone to neglect the taking of proper exercise. The neglect of this hygienic requirement is sure to result in ill health. It is a requisition that cannot be dispensed with. It is a part of the constituted condition of existence, emphasized in *holy writ*, and whose verity is a continual matter of experienced observation. Nature has no use for the incorrigibly slothful. Many persons are too lazy to live; and they do not live. Their lives are shortened—disease fastening upon their ill-nourished bodies, sweeping them from mundane existence. The exercise of the most important parts of the body—the internal nutritive organs, comprising the heart, lungs, stomach, liver, etc.—naturally should receive the first consideration in hygienic living; yet these are the parts that are usually most neglected by the class of fair beings mentioned, as well as by most persons of sedentary habits and occupations. The consequence of such neglect is an inactive condition of these parts, which causes a poor circulation of the blood, poor digestion and assimilation of the food eaten, and a deficient amount of excretion of the worn-out nutritive elements, which collect in the general system, contaminating it, producing headache, nausea, sick, creepy sensations, pain in the bones, bad breath, a thick, muddy complexion, dull, yellow eyes, and a general depressed condition, that represses mental activity and all healthy cheerful life. All these undesirable results would not occur if these important parts of the body were properly exercised. There are many serious impediments to the taking of this exercise, among which may be mentioned the discomforts incidental to too hot or too cold temperatures; or to very stormy, wet, or dusty conditions; and also the restrictions of modern fashionable dress and customs. A profound and prolonged consideration of these facts has led to the invention of a very ingenious, yet comparatively simple mechanism, the moderate use of which affords the amount of exercise necessary for the preservation of vigorous general health, and the cure of many disordered conditions that may have arisen from its neglect. Furthermore, this exercise can be taken in the most convenient, comfortable, effective, and expeditious manner—more so than by any other method hitherto known. This mechanism is known as **THE HEALTH JOLTING CHAIR**, an illustration of which is here given. The chair affords an exercise similar to that given by a saddle-horse. It can be regulated so as to give it in any degree of severity desired; strengthening and increasing the activity of all the organs named, and beautifully developing the arms, shoulders, and chest. We would be pleased to send free to any address an interesting pamphlet entitled "Exercise of the Internal Organs of the Body Necessary to Health." It contains a full exposition of the subject, including a description of The Health Jolting Chair and the remarkable benefits that are derived from its use.

The Health Jolting Chair Company, 150 West 23d St., New York.

Jolt Your Wife to Life!: Before vibrating chairs lulled and tickled women into bliss, the health jolting chair (left) would temporarily make her forget her "hysteria." Remember: "Nature has no use for the incorrigibly slothful."

HUMAN GUINEA PIGS
KILL THEM TO SAVE THEM

The most notorious experiments on humans were led by Nazi Heinrich Himmler. He was convinced that he could find a way to revive the dead. When staff surgeon of the Luftwaffe and SS Sigmund Rascher dissected bodies and found the hearts still beating after the chests and skulls had been opened, the Nazis thought they had discovered a way to revive the dead.

One disturbing test that Himmler ordered Rascher to conduct was to quick-freeze his patients—the Jewish prisoners—and then revive them. Rascher used hot water, arguing that it was the fastest way to get heat to the body, but Himmler recalled German folk tales of herbal teas taken by frozen fishermen, and of their wives taking them to bed to revive them. Naked German guards were ordered to revive the frozen corpses as Nazi scientists scribbled notes while watching the necrophilia.

The U. S. wasn't innocent of medical testing on patients, either. The Tuskagee Syphilis Study in Mississippi in 1932 on 400 black males was one of the most notorious. In the 1960s, approximately 20,000 prisoners participated in "voluntary" medical experiments, including inmates in Ohio who were injected with leukemia and then given various medicines to try and treat it.

In 1953 at the height of the Cold War, the U. S. Army wanted to test the dispersal of aerosols and fallout on a city. Four sites in Minneapolis were chosen for the eighteen-month study with sixty-one releases of a toxic chemical. According to *The New York Times* of June 10, 1994, "...the Army sprayed clouds of toxic material over Minneapolis dozens of times and may have caused miscarriages and stillbirths, a public television station reported. The spray-ings in Minneapolis and other cities were described then as part of an effort to develop an aerosol screen to protect Americans from fallout in case of an atomic attack.... One of the sites sprayed in Minneapolis was a public elementary school where former students have reported an unusual number of stillbirths and miscarriages." The substance sprayed from rooftops, out of the back of open trucks, or from mysterious boxes at intersections, was revealed to be cadmium sulfide, which can cause lung damage, fatty degeneration of the liver, and acute kidney inflammation.

In the early 1960s, the infamous drug thalidomide was prescribed to stressed mothers as a general sedative. Made by the company Chemie Gruenenthal to help women calm down from the pressures of pregnancy, in forty-six countries thalidomide caused 8,000 seriously deformed and handicapped kids to be born. The drug was finally removed from the market.

More recently there has been a rush to reap the benefits of biotechnology and ethnopharmacology. Biotech labs routinely create tens of thousands of novel strains of bacteria or yeasts, culling them and patenting any that might be moneymakers. The rest are routinely flushed down the drain. Ethnopharmacologists study with traditional healers the world over to find (and patent) plant-based medicines. The anti-inflammatory drugs Vioxx®, Celebrex®, and Bextra® have all recently been withdrawn from the market, and the antidepressants are starting to show their fangs as well. The dream of a pain-free life continues to recede, and patients are again realizing that drugs may not solve their problems.

Visions of Rockets: According to this lad's view (above), every guy gets a girl and a rocket.

Blasting Off: The eternal dream of launching oneself into the air (right) is realized in this personal rocket. See what you and a couple pals can do to get to space; getting home is another project.

Space Colonies
SLINGSHOT TO THE MOON

DREAMS OF FLYING
THE FASTEST MAN ALIVE

Humans have been fascinated by movement ever since we looked up to see birds flying. (Or were UFOs the inspiration?) Maybe it was the first time a highly intelligent human was chased by hungry lions and longed desperately to be elsewhere, or at least to be moving sixty miles per hour surrounded by a metal and glass dome. The last thoughts of that unlucky ancestor might have been, "Gee, that could make this little chase fun; I could even make some money selling such a ride."

For our soon-to-be-consumed forefather, the greatest velocity ever achieved was about twelve miles per hour. Yes even our obese, out-of-shape bodies can run twelve miles an hour in a panicked rush. The fastest a human traveled before the invention of the rocket sled was the speed of a horse, more properly, the speed of a panicked horse about to be overtaken by a predator. Thirty-five miles an hour, let's say. A Geo *Prizm* takes about four seconds to surpass that speed, less on a well-sloped freeway on-ramp.

Not to dwell on the macabre, the only time early humans traveled as fast as we habitually go on the freeway, they were most probably in free fall, victim of some awful accident that propelled them off a cliff or out of a tree. You can surpass 140 miles an hour given a fall of 100 feet or so. Alas, that human speed record was held only by those buried at the base of cliffs. Guess what they were thinking as they fell? Besides the real thrill of the ride, and the realization that they could make a ton of money if they could package it, I bet some sort of rocket pack might have flashed before their minds. Dreamlike, they imagined themselves flying, without wings, just blasting away from the mortal coil.

As Thomas Edison quipped, "It is failure that is the mother of invention."

A GATE TO THE UNKNOWN
TAMING UNIMAGINABLE FORCES

Gunpowder was the first rocket propellant, used as a pretty diversion until it was put to use to expel Mongolians trying to breach the Great Wall. From the tenth century, gunpowder came to be part of the worldwide quest for domination, and was one of the main factors in the conquest of the Americas.

A balloon released to zip around is a rocket. Superheated water works as well, blasting into steam as it escapes the pressure of a rocket chamber. But the modern rocket uses a liquid fuel, exploding as it compresses through an exit nozzle. Our bottle rockets use a stick as a stabilizer (maybe an eighty-foot-long stick would make jet packs possible); letting the rocket spin (or gimbal) was invented in 1844, reducing the need for the stick and creating a new age of

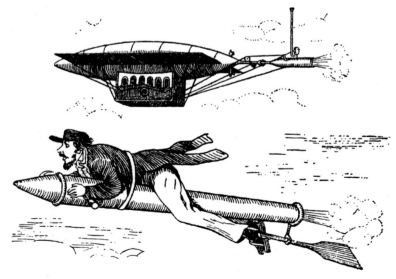

Steam Rockets: The explosion of water expanding into steam is perfect for these rocket designs from the 1860s (above). Of course, there needs to be a fuel source for heat, or incredibly high pressure and an extremely sophisticated nozzle. Gunpowder or gasoline proved much easier, especially for this lookalike precursor to riding a nuke in *Dr. Strangelove*.

From the Earth to the Moon: Gunpowder and a huge cannon propel this ship like a bullet (left). G-forces were unaccounted for by space fiction pioneer Jules Verne in his seminal masterpiece *From the Earth to the Moon*.

"It has often proved true that the dream of yesterday is the hope of today and the reality of tomorrow."

—Robert H. Goddard
(1882–1945),
pioneer of
liquid-fueled rocketry

The Flying Barrel: Okay, so it's a jet, not a rocket (left). The "annular winged plane" took off vertically, then flew horizontally—in theory.

rocket accuracy. In 1917, Robert H. Goddard added a supersonic nozzle to the combustion chamber, blasting fuel into the burning rocket stream and multiplying the force and the heat of the rocket and increasing its efficiency.

MAD SCIENTISTS
THE SLINGSHOT EFFECT

The mental leap to spaceflight was inspired by the writings of Jules Verne, who wrote of a trip to the moon in ships blasted from a cannon. A Russian teacher named Konstantin Tsiolkovsky, deaf and unaware of the innovations going on in the world, conceived of a "reaction machine" that would burn liquid fuel to propel a rocketship and bring humans to outer space. He calculated precisely how much power was needed to launch a rocket into space, including the mass of the payload and the diminishing mass of the fuel; his equations are still used today.

Both Tsiolkovsky and the American Goddard initially imagined a great spinning machine that would throw a ship across the horizon and into space. Rocketry was such a far-out science that Goddard kept his passion to himself until his late twenties when he received a patent for a liquid fuel "rocket apparatus." Goddard was quite successful with his work, and developed a recoilless rocket launcher for the war effort by 1917. In 1920 when one of his scientific papers hit the popular press, Goddard saw the headline in *The San Francisco Examiner*, "Savant Invents Rocket which will Hit the Moon" and his fears of being labeled "mad scientist" had come true.

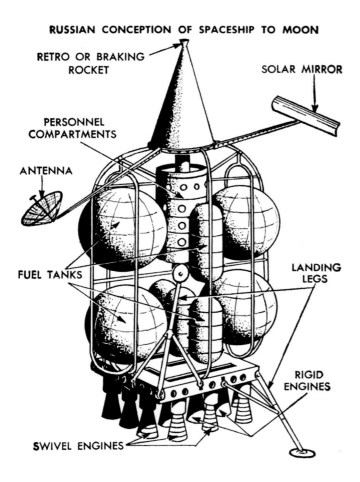

RUSSIAN CONCEPTION OF SPACESHIP TO MOON

RETRO OR BRAKING ROCKET

SOLAR MIRROR

PERSONNEL COMPARTMENTS

ANTENNA

FUEL TANKS

LANDING LEGS

RIGID ENGINES

SWIVEL ENGINES

A Cutaway: Takes some engineering to put one of these together (above). Note the dozens of rocket nozzles and the wings to control yaw, pitch, and attitude.

The Shuttle: This early Russian rocket (far left) concept ferries a winged spaceship that could glide back to Earth. Note the wings to stabilize.

Nice Little Plane on Top of a Big, Big Rocket: Imagine re-entry in this tiny glider (right). Ouch. This makes the space shuttle look like it's from the twenty-fifth century.

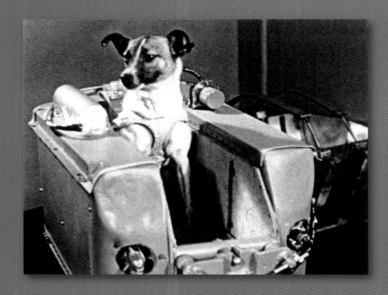

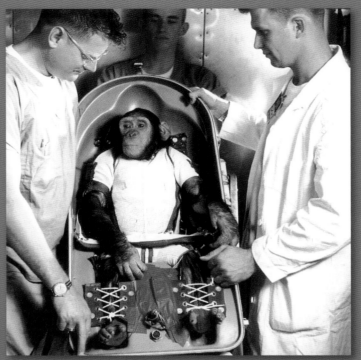

Underdog and Wonderchimp: There is still no proof that humans can survive long outside Earth's magnetosphere. Surviving in the almost vacuum of low orbit was studied first by smaller animals, like the Russian's Laika and NASA's Ham (top left and right).

The Slingshot: The two great inventors of rocketry, Goddard and Tsiolkovsky, first imagined a slingshot to space (bottom left). Feel the centrifugal force pulling at this prospective spaceman on this sci-fi Tilt-a-Whirl?

Those dastardly Nazis were the first to perfect a rocket missile that could reach orbit, the *V2* (vengeance weapon 2). With a range of 150 miles, it could hit a city-sized target and so was more a weapon of terror than one that could strategically be used on the battlefield. The scientists who perfected the *V2* went on to work in the U. S. and the U. S. S. R. after the war, immersed in the Cold War space race.

FTL: FASTER THAN LIGHT
BRINGING STAR SYSTEMS CLOSER

The rocket is standard fare for futurist follies. Rocket packs, sleds (on rails), cars, boats, and airplanes have all been proposed, and of course, tested. To achieve orbit, a speed of mach twenty-five (twenty-five times the speed of sound, 750 miles per hour) is needed, and speed is what rockets are good at.

In space there is no air, so propulsion engines need to bring their own fuel, another reason for choosing rockets. Of course if you understand the immensity of space you may know that mach twenty-five won't get you far. Warp speed is needed.

Warp is the speed of light. If you've been paying attention, the speed of light is supposed to be the fastest you can go, an uncrossable frontier. But if you wanted to travel to Sirius, the most significant nearby star, it would take you ten years to get there if you could go as fast as light, and ten years to get back. Of course relativity says you'd age less than twenty years, but still it makes for a big mess. And the center of the galaxy is 33,000 light years away.

Faster Than Light (FTL) drive is simply necessary for making space travel interesting enough to write about. The reality may be a bit more tedious. Theories of FTL travel include "warping" through hyperspace, transdimensional jumps,

and psychic travel. The ram jet uses a slingshot effect to leave solar orbit, then scoops the sparse elements from the almost vacuum of space into its fission jets; acceleration amplifies the amount of fuel it can scoop and the speed it can achieve. If it can be imagined, it can be achieved, right?

REACHING MARS
BUXOM ALIEN BABES

Mars is the prime threat of alien invasion, maybe since the long ago night when it was named for the god of war. It was Martians who invaded in H. G. Wells' 1898 *War of the Worlds*, and really made a splash on Halloween 1938 when the radio drama version was broadcast.

In Donald H. Menzel's book *Flying Saucers* from 1953, he debunks UFO sightings and saucer sightings as poppycock, but he writes that the canals of Mars are surely formed by water and the vegetation is nearly visible by telescope. He hails a sci-fi writer as a visionary of Mars' inhabitants. "To my mind, the best model of a Martian was that suggested many years ago by Hugo Gernsback…(and his description of) a male Martian, whose characteristics indicate the general trend of evolution on the planet."

It was just a matter of time before we'd have exotic aliens—especially Martians—as mates. Gernsback described Martians in *QUIP* magazine from 1949 as "…almost ten feet tall, their huge barrel-shaped upper bodies surmounted by a tremendously ponderous head, shell-like ears almost a foot across, and an elephantine nose three feet long. Even more impressive are the stalk-eyes, which project out of their heads and can telescope in or out…at the top of the head we discover two huge insect like antennas—the telepathic organs, which the Martian uses for communication with his fellows."

Gernsback's view of women—at least earthly women—was clear in his description of the red planet's weaker sex. "The Martian female is about six inches shorter than the male. Her waist is somewhat more shapely, but her outstanding characteristic is the possession of double antennas.… This characteristic enables her to generally out-talk the male and otherwise confuse him. Thus, on Mars, the real rulers are the females."

"Mission to Mars" was a Disneyland attraction, but is also a theme for many of the world's space programs. NASA's Mars rovers have broadcast a vast amount of data from the planet, though the presence of water remains theoretical.

The Canals of Mars: A male Martian (below) singing to the moon, according to Hugo Gernsback.

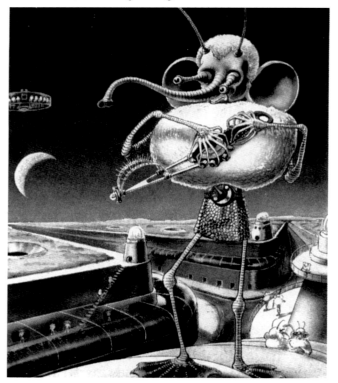

Rockets Gone Wild: Little red bugs, wide-finned firecrackers, racers, and big sport utility rockets (right). So when am I gonna get mine?

"Captain Requests Permission to Blast Off:" Look at all those plebes riding these rockets (bottom). Oh to be the captain in that amazing uniform; with one wave of your wand you're off.

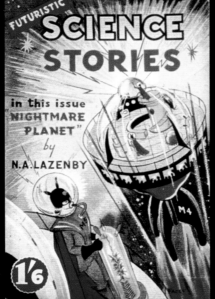

"[Before man reaches the moon] your mail will be delivered within hours from New York to California, to England, to India, or to Australia by guided missiles.... We stand on the threshold of rocket mail."

—Arthur E. Summerfield (1899–1972), U. S. Postmaster General

Multistage Rocket: Impressive artwork shows the winged shuttle blasting away from its lift stage, and the heat of re-entry blasting around the stage (below). Collier's had several full-color covers featuring the space program in the early 1950s, from the moon lander to turreted escape pods holding dozens of astronauts. Very realistic, and very close to the actual space program to come.

The Forevertron

JULES VERNE'S ROCKET TO THE MOON

Tom Every of Baraboo, Wisconsin, is building a trans-temporal spaceship in his backyard. Under the six-story-tall *Forevertron*, the eccentric Every explains his experimental machine: "Two people can sit there in the 'Celestial Listening Ears' and listen for alien voices from the heavens and beyond. Information will then be relayed to the 'Overlord Control Center.'" We're walking around the enormous electro-magnetic space station he's building to propel himself to the heavens.

"I started working on the *Forevertron* in 1983, and by God you do something until you drop over dead! It now weighs 400 tons, and all the things on it come from the American Industrial Revolution. It's from around the 1890s, so it has a Victorian aura to it." In fact, Dr. Evermor's—as Every often refers to himself in the third person—creation derives part from the mentality of turn-of-the-century explorer and part mad scientist, like Captain Nemo meets Nikola Tesla. "If you like Jules Verne, you'll love talking to me about the possibility of what could be.

"See that? That's the 'Gravitron' where the good doctor"—referring to his alter ego—"will de-water himself to reduce his weight before blast-off. Almost all of human weight is water, so this will make the blast-off easier." The pod-like Gravitron has electrical gizmos—from transformers to dynamos built by Thomas Edison—sticking out in every direction.

The last stop is a new shipment of enormous I-beams that have just been delivered next to a section of the NASA *Apollo* decontamination chambers that he scrapped. "Next summer, I'm going to finish up the Overlord Master Control, which will be the biggest section so far and connect over to the 'Juicer Bug' to give us extra juice."

Doctor Evermor then describes the scene of the much-anticipated day: "Then the doctor will walk down that spiral staircase, go over the bridge, and step into the copper egg inside the glass ball before being shot into space. A hum buzzes through the air, all the lightning rises, and the doctor climbs into his pod. The lights go from red to amber to green. They pull all the switches. The good doctor in his trans-temporal copper egg chamber shoots out on his magnetic lightning force beam. All the non-believers and doubting Thomases are drinking tea. It's a very happy day and everyone in the band begins playing. Everything is built for that great day."

If the blueprints for his *Forevertron* masterpiece are ever completed, perhaps it won't take him into the heavens, but his life's work would at least make it the largest scrap metal sculpture in the world. "If it could be, why not make it be?" Dr. Evermor asks.

Wheels and Sleds: Gleaming outriggered ships (above) skating around home port, slim needles and the wheels they puncture.

WHEELS IN THE SKY
MEET YOU IN ORBIT

"Earthrise" is the name of the photo. In 1968, *Apollo 8* orbited the Moon, and watched as the Earth rose from behind the Moon. The delicacy of this gem in the void of space has inspired a renewed quest: to protect the legacy of Earth by moving into space.

There have been short-lived space stations since that time, from *Soyuz* to *Skylab*. While *Mir* lasted fifteen years, the *International Space Station* is greater in size and in research power and appears poised to last (or is that another folly?). The great question is, if some can leave Earth, does that make those on the planet more secure, or less? Until the 1970s, space was considered a weapon-free zone, but the U. S. government's Star Wars program has threatened an arms race bringing the worst of human endeavors into the heavens.

The great wheels in the sky that we know from *2001: A Space Odyssey* are the next generation of space city. Why a wheel? Well, the hub is relatively motionless and great for docking spaceships, and the rim has an artificial gravity created by the spin. On a grander scale a cylinder would do the same, a tube spinning to make centrifugal gravity on the inside of its rim. Or a tube shaped like a ring. Or a swarm of satellites ringing a planet. *Ringworld* is the name of a novel by Larry Niven in which a huge belt-like ring orbits its star, with more surface area than a million Earths. Bigger yet is the Dyson Sphere with a star at its center and an immense amount of stellar energy available for all the needs of modern life.

Some of these are based in physics and others on a desperate sense that we need a place to retreat to, but either way the sky is no limit.

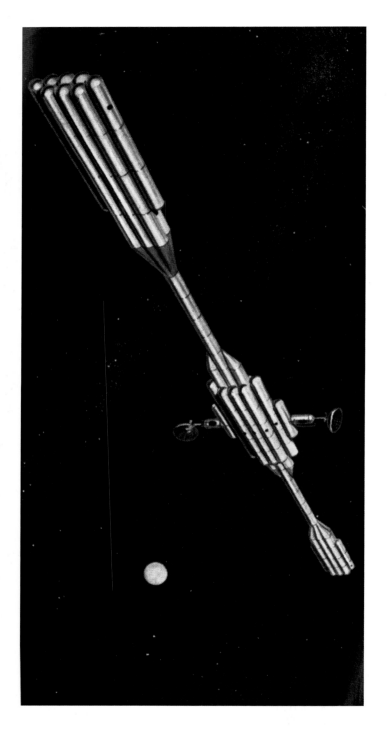

INFINITE POWER
MATTER-ANTIMATTER BATTERY PACKS

The promises of unlimited energy mean that even the tiniest inconvenience can be banished from human life, the last itch assuaged, the final chink in our perfection stuffed. Nuclear energy will be replaced by matter-antimatter implosions. The energy simply must be channeled correctly, into the web of necessity in finer and finer pathways. A city draped in energy, to keep the weather or enemy away; a shield of energy to take into battle or to take on a 100-mile-an-hour ride; a suit of energy to protect the skin, to amplify the senses or the muscles and accommodate the organic body; an inner field of energy to cleanse the body, to protect it from the smallest invaders, to amplify the metabolism and accommodate the spiritual being.

The stars have been imagined to be the souls of the dead, the Milky Way called the spine of the night, the moon a mirror in the sky. The last thing we need, as students of history and science, is to discount these myths as absurd and ignorant fantasies. Instead we can see the truths they hold, and remind ourselves that our own theories of the laws of nature are but building blocks, and yes if placed well they can be a great foundation, but we know not where they lead. Our own calculus is but a metaphor, a very useful one, but still just a play of light on the surface of the depths.

Representing the Not-Yet-Imagined: Many choose to overwhelm the audience with details and complexity for things yet to come. In this image (left) we see the alien-ness of the future through bizarre simplicity.

VANGUARD
science fiction

FIRST ISSUE: KORNBLUTH · JONES · DEL REY · DE CAMP

SOS: PLANET UNKNOWN
By A. Bertram Chandler

EDITED BY JAMES BLISH

The Dangers of Rocketry: For all the glamour of rocket science (above), let us never forget to look in the rearview mirror before engaging the engines.

International Tourist Wheel: This wheel has its windows in the floor and a physically impossible form of propulsion (right): blasting the rockets at the hub would spin the wheel end over end. Is that rocket a pleasure cruise? The space walk seems purely for fun.

Be a Pig!: With futuristic technology, merging man and beast will be possible (center), but who would want it? Think about the carnie shows in 2050! No longer will it be a trick of the light as in this 1815 caricature.

The PIG-FACED LADY, or Manchester-Square

Fast Forward to 2050

FUTURISTS PREDICT FOR THE NEXT GENERATION

ANTICIPATING THE FUTURE

FLASH FORWARD FIFTY YEARS

Immortality? Implanted cell phones? Crash-free cars? Glass houses? These are predictions tossed around by a group of present-day oracles, meeting monthly to foretell the future. Hank Lederer, a nanotech aficionado and outspoken advocate of scientific optimism, has agreed to reveal what the Minnesota Futurists envision as possible advancements within the next fifty years.

The Minnesota Futurists prefer to call themselves "anticipatory thinkers" rather than visionaries or psychics who predict the future. The head of the group, Earl Joseph, who worked twenty years as a professional futurist for Unisys, placed a crystal ball in his foyer as a spoof of this hocus-pocus. He points out that they study anticipatory sciences, extrapolating past and present trends into the future. In other words, if you can imagine something (and track its potential to happen) it can come true.

What's the difference between anticipating and predicting the future? Hank Lederer explains, "There are so many variables that you don't know what the hell is going to

Redesign Your Body: Just as early sci-fi promised to build bodies from the skeleton on up (above), modern futurists predict personal redesign to create a new look for the old you. Sure, this may not be building your own private personal bombshell (Why didn't Dr. Frankenstein make a fraulein?), but at least you can sport the most perfect nose job ever conceived.

happen. That's when a leader or a group comes in and says what they want to see happen." Presumably futurists are in a good position to make helpful suggestions.

From this plotting, he hopes to predict all the great (and not-so-great) things that await us in the future, from cryogenics to extraterrestrial colonies. "In the short run we expect too much change, and in the long term we don't expect enough," Lederer said as a caveat that predictions can either be flat-out wrong or not go nearly far enough. He noted that almost no one predicted how the Internet would change our society.

"Advances in science in the future all depend on a relative level of civilization. It's based on the hope that fundamentalism won't take over the world. If we don't do stem-cell research, some other country will do it." Lederer views "religious zealotry derailing science" as a major hurdle to positive advances. He warns about faith-based science and the idea that, "'We don't look because we're not supposed to know,' which is very anti-science."

The other caveat is war and terrorism. Lederer considers himself a Pollyanna, envisioning the fantastic advances of which humans are capable, rather than a dystopian pessimist. Even so, "A few of us worry that there will be an attack that will make 9/11 look like a bake sale. We search only about 6 percent of container ships while Hong Kong does 100 percent. A dirty bomb on a container could spread the radiation across the country.... This will look like the good ol' days. There are lots of doomsday scenarios out there. I love high tech; it's people that are no good!"

Nevertheless, here are the optimistic visions of tomorrow when present trends are extrapolated twenty-five years. As Hank Lederer pointed out, "[Futurism is] not supernatural or magic or intelligent design or whatever. It's just science."

TECHNOLOGY

2010 TO 2015: RADIO FREQUENCY IDENTIFICATION (RFID)

"Wal-Mart's into this already and using it on pilot programs, but it's still too expensive at ten cents per chip. RFID will take the place of bar coding. The passive RFID chips cost a dime for the tiny little chip that has a sequenced bar code in it. Like a transponder on an airplane that reflects radar in a big way, it reflects the signal from the radar and tells what it is. You'll walk through the door and it'll automatically charge you. Even when you buy an apple, it'll have an RFID chip, so it'll have to be able to be digested.

"By 2010, you'll start to see them in check-out aisles, but by 2015 it'll be everywhere and replace bar coding. Active RFID chips will actually transmit a signal. They'll be implantable. GPS systems will read everyone so your kids can't be kidnapped and a wife can always tell where her husband is. And who he's with! If we're worried about losing privacy, this is going to be huge. People are going to continue to sacrifice privacy for security. The more technology and the more people, the less privacy and therefore less freedom. Because you have to keep track of them. Once again, it's people that cause the trouble!"

2012 TO 2015: FOLDOUT COMPUTER SCREENS

"Nanoscale material used for flat-panel displays. Flex displays [will be able to] fold out from your phone or computer." These expandable screens, and the virtual screens that may come someday, are already standard fare on sci-fi TV.

2018: NANOTUBES

These microscopic building blocks are 100 percent carbon and enormously strong. "An early consumer product to use nanotubes was Lance Armstrong's handlebars. [Nanotubes] are strong like diamonds and now cost $1,000 a gram."

The current problem is attaching them into a usable string. A nanocable would be, "…super sharp because it's only four microns thick." A nanocable the size of a human hair "could hold a 3,000-pound car. If you used it in construction, suspension bridges would be super light weight, but [the braces] can't be too thin because birds might hit it [and be sliced in two]." Put just 2 percent of nanotubes in a carbon mix and a mile-high skyscraper could finally be realized.

2020: IMPLANTABLE CELL PHONES

"You'll have two types of cell phones. One can be implanted under the skin." Will anyone actually do this? "Oh yeah, they'll do it. They have it behind their ear, and then people will think why not just implant it? You'll have an answering machine in there that'll be so smart that it'll know when you don't want to talk to someone and maybe you'll get back to them someday. It's gonna seem like telepathy. Two people on the opposite sides of the room can talk to each other," and barely look like they're saying anything because, "…they're only sub-vocalizing. This is going to be very popular." Though we'll have to learn how to talk by sub-vocalizing, silently, without moving our mouths. Will we still gesture with our hands while sub-vocalizing?

"The other kind of cell phone will be a big screen that you can unfold. The other option is just to have it on glasses that project it on to your retina. They're going to have so many different kinds depending on what people feel comfortable using. All this is based on very, very tiny circuits."

2030: SMART PAPER

"This is electronic paper that is like a liquid crystal display that you plug in on the side—or maybe it's battery or solar powered. It would look like a printed page and you could press on the corner and it would just go to the next page. The computerized display would look like paper and you could fold it up. Your newspaper would be just one sheet."

MEDICAL BREAKTHROUGHS

2010: STEM CELL THERAPIES

Tissue will be repaired by, for example, injecting stem cells into an organ and have it regenerate itself. The roadblock is a supposedly ethical one. Hank Lederer reiterated the argument that has slowed research: "Destroying the embryos is OK, but you can't do experiments on them because they could become a baby someday." Ways around this are to "take a skin cell, take out the DNA, put it in a woman's egg cell that has the DNA removed. Give the cell a shock so it will forget it was a skin cell. Then take out a gene and have an unimplantable 'un-embryo.'" And the availability of stem-cell therapies? "Maybe not here [in the U. S.], but you can go somewhere else—Mexico or Korea."

2012: SOME CANCERS CURED

Various technologies are promising, including the technology of a University of Michigan researcher who developed a "multi-layer nanosphere [medicine] that is forty nanometers in diameter. It can go right through a cell's wall without even raising its defenses. It's a miniature poison capsule that would probably be inhaled. There are three layers [of the medicine]: the first layer breaks away once it's in a cell; the second layer looks for too many telomerase in a cell, which indicates a cancer cell (except for in hair cells); then [the third layer] finally releases its chemo poison only in cancer cells."

2015: CUSTOMIZED DRUGS

Pharmacists and doctors will "develop drugs that are custom designed [for each patient] that would be fatal to one person but can cure another."

2015 OR 2020: ARTERIAL REPAIR

"Now we have a little camera that is the size of a capsule that you swallow and it sends little radio signals with good live action images. If they could make it even smaller to go through your arteries and capillary system—it would have to be the size of a red blood cell—then it could have a little grabber that could cut away plaque. Maybe it could have a vibration system that breaks down the plaque but not the artery walls. This is the holy grail of nanotechnology."

2015: LIFE EXTENSION

"When are we going to extend our life spans more than one year per year?" As we start curing cancer, beating Alzheimer's, etc., we'll "maybe start expanding the average life span by more than one year per year beginning in 2015."

2020 OR 2025: IMPLANTABLE ORGANS AND LIMBS

"Grow a new hand or organ or new eye somewhere and then implant it or else just grow it right off of your body. Like a salamander grows a new tail with bone and tissue, stem cells can grow into a new limb. The problem is how to attach it to the nerves and the brain."

2040: IMMORTALITY

"It's called 'Extreme Life Expansion' and you can basically choose how long you want to live. We'll cure the disease of aging, but this will create a huge population explosion. If everyone is on the planet for three times as long, you have three times as many people. I'm afraid it might be only for the very wealthy, though.

"In the future, virus-sized computers will be able to go inside you and change your genes. Nanotechnology computerized systems will run through your body and check for bad DNA. You'll need a computer to doctor up your genes so you won't get old.

"People will be able to live as long as they want, or at least until they get bored or have an accident. I don't think most people will want to live more than 200 years, though. Many will probably just get bored and kill themselves, but that's my opinion."

2045: PERSONAL REDESIGN

"Humans are composed of 10 trillion cells with all this DNA that's all the same. If each cell never gets old because they all think they're just twenty years old, the enzyme that repairs DNA could fix all the cells. You could also change the way you look. If there's a particular actor or actress who is hot, you could look like them. You have to buy the software, but you can have very subtle changes. Change your appearance, your eye color, hair color, or whatever." Somebody might say, "I want to see what it's like to be a woman for a while. If you want to be a lion or a tiger, you could probably do that, but that's way out. This all assumes that science will continue as it has been going, of course."

HOUSES

2010 TO 2030: SMART GLASS

"Soon we'll have self-cleaning glass. Pittsburgh Plate Glass is working on this right now. Now they have semi-smart glass that uses sunlight to repel dust." With smarter glass, "you could write on there [the glass] with a felt pen and in ten minutes it'd be clean. Extra-smart glass will repel heat or darkness when you want to. This glass will let sunlight in during the winter but keep it out in the summer.

"By 2020, we'll have really smart glass, powered by electricity. If someone drives by with their stereo blasting, it'll cancel out that sound." This will use the same technology as in active noise-canceling headphones available now. This smart glass "won't kill a bird if it hits it, but it'll give. Or if a baseball hits it, it won't crack.

"In 2030, it'll look like whatever you want it to look like. If you want the ocean to look at or a forest, when you walk around it'll change. You can have a sunset that'll change as you walk by it. Also, it can turn into a TV set. The next step is that the glass becomes a window or wall, or a door, when you want it to."

2016: SMART ROOFS

Rather than wasting all that sun beating down on your roof, and short of installing expensive solar panels, scientists will develop a "smart material roofing that is solar collecting using nanoscale science."

2060: HOUSES GROWN FROM SEED

With advanced nanotech, "you'll be able to grow your own house from air and water with a seed. It'll grow as fast as thermodynamics lets it. You can grow a good-sized house in two or three years. Doors, windows, floors, appliances will need elements beyond carbon [from air] and hydrogen [from water]."

TRANSPORTATION

2012: ACCIDENT AVOIDANCE SYSTEMS

"Mercedes already has a system on a truck with the brakes and gas and steering wheel hooked up to the computer so you can't run it into a brick wall. Right now you can buy adaptive cruise control, so you can stay right along with the flow of traffic [only up to ten miles per hour]. This uses lasers and radar to match your speed to the car in front of you. This costs about $2,000 [and is available on] BMW, Mercedes, and Lexus." By 2012, it'll be as standard as seat belts are now.

2015: AUTOMATIC AUTOS

Using intelligent agents, essentially computer programs that can learn, cars will be completely automatic and not require a driver. "The ultimate intelligent agent will bring me here…and I'll just sit back and read the paper while it drives. I want it to have emotions, or rather to understand emotions. Then to shut up if I'm in a bad mood. It's software that can learn and be your friend.

"Self-driving cars won't need smart highways. The cars will use video, radar, laser, infrared, and mostly GPS, which will take it within about three feet of where it needs to be and then the other sensors will bring it the rest of the way." The self-driving cars will "react within a millionth of a second—thousands of times faster than we can react. These self-driving

Raygun Gothic

FLYING SAUCER OVER L. A.

On October 4, 1957, a little "beep-beep" was heard orbiting Earth, and the U. S. panicked. The Soviet Union had launched the first artificial satellite, *Sputnik*, and the 184-pound spiky ball transmitted nothing more than its beeping, which could be picked up on radios across the U. S. Death rays and missile attacks couldn't be far behind.

President Dwight D. Eisenhower demanded that the U. S. catch up with the Russkies at all costs; the Space Age was born. While the government formed NASA and pumped money into schools to support science education, architects jumped on the space craze and built bold new buildings based on sci-fi dreams. Diners, motels, casinos, and bowling alleys raised bold neon signs with stars, satellites, boomerangs, and rockets. Southern California, already famous for its "Googie" or "Populuxe" architecture, embraced the new space-age style that cyberpunk writers later dubbed "Raygun Gothic."

As part of the $50-million "Los Angeles Jet Age Terminal Construction" project, the "Theme Building" with its spider-like legs became the epitome of the new futuristic architecture. Erected in August of 1961 at a cost of $2.2 million, the white saucer was held by 135-foot-high parabolic arches made of 900 tons of steel. The windowed disc in the center holds the Encounter Restaurant, which is loaded with smooth kidney bean shapes and smooth plastic design later adopted by *The Jetsons*.

This pinnacle of space-age kitsch was trumped the very next year by the Seattle Space Needle, raised as the centerpiece of the 1962 World's Fair with the forward-looking name: "Century 21."

cars will be enormously safe. It'll know the fastest way to get there and avoid any storm damage or roadblocks."

2030: SELF-REPAIRING ROADS

To save billions on road repair, microscopic intelligent agents (IA) will be installed in the interstate system. "They will replace the roads with smart roads, which will have microscopic computers in them that will automatically repair themselves. This is far-out stuff."

2030: HYDROGEN AUTOMOBILES

The problem with hydrogen cars is that they "require a new trillion-dollar infrastructure." Better technology is available. "For cars, I think it's unnecessary."

ENERGY

2015: MORE WINDMILLS AND HIGH-EFFICIENCY SOLAR PANELS

Right now, a 3,000-foot-tall solar and wind-power tower is planned to be erected in Australia for $380 million, and it will produce 200 megawatts of power. "If they could only make those windmills prettier."

Windows, using smart glass, will double as solar panels. Roads could also be transformed into solar panels to give them the energy to repair themselves.

2020: SAFER NUCLEAR POWER PLANTS

We'll use "newer, smaller, safer nuclear power plants and get rid of the oil and coal." We'll "reuse the radioactive material and only have 10 percent of waste that we have now." We can "shoot old nuclear waste into the sun, but it'd have to be very safe."

2020: DISTRIBUTED ENERGY PRODUCTION

We'll use "wind farms, solar, and fuel-cell combination for storage [of energy], which means we won't have to have big power plants. In the summer, the hotter it is the more energy we'll get from solar panels."

ENVIRONMENT

2015: CLEAN WATER

"Right now, a guy has a super straw that has a nanoscale filter system about a foot long. You could put it in a dirty puddle or a swamp and the water would come out clean. [It doesn't work for salt water.] When the filter is full up, it just stops."

The problem is removing salt. "Super filters using nano-tech, can do everything but salt. Only evaporation or reverse osmosis gets rid of the salt, but solar panels will make desalinization cheaper."

2040: CREATING CLOUDS

"Global warming is not going to slow down. With nanotech, however, you could create more clouds to reflect sunlight so Earth doesn't warm up so much. By 2040 or so, we'll be able to juggle the environment and eliminate global warming and start reducing pollution enormously."

FOOD SUPPLY

2020: WASTE- AND EMISSION-FREE AGRICULTURE

Greenhouses covered with plastic film over a nanotube structure will be used to raise almost all food. "It'll be cheaper to have one of these water recycling greenhouses than all that diesel irrigation."

2050: FOOD CREATOR

"A little box the size of a microwave will be your food builder, or food creator. It will take air, water, and sunlight and turn them into meat and milk or Brussels sprouts, whatever you like. It'll cook it, too. Every two hours you can get a glass of milk out of it and every six hours you can get a hamburger.

"Where do you get these things from? The top is all solar panels and grills on the side get air. The bottom is open to the dirt and sends little microscopic tubes down into the ground sometimes hundreds of feet down" to get water and necessary minerals and other nutrients. After all, "the meat and milk in a cow come from the grasses that themselves come from air and water, right?

"The first one will cost about a billion dollars, but it'll come down from there. The other holy grail of science is to recreate photosynthesis."

DEMOGRAPHICS

2030: COMPUTER AND GENETIC WARFARE

"People used to worry about NBC (nuclear, biological, and chemical) warfare. In the future, it'll be more GNR (genetic engineering [genomics], nanotechnology, and robotics). Bill Joy wrote an article in 2000 in *Wired* that it's going to kill us all. You could figure out a super virus that would only get people with a certain genetic make-up: ethnic cleansing. Or a super virus that goes into the brain to make everyone docile, as a sort of a tranquilizer. Are people smart enough to really solve their problems? So far they have been, more or less. Or are we like kids in a pool of gasoline playing with matches? As soon as we drop it, we're all dead."

2060: POPULATION PEAKS

Even though the birth rate will decline, "population will grow anyway because of aging. It'll peak out around 12 billion in 2060. Half of the people won't want to live longer because either it's not natural or against their religion."

2060: POVERTY ELIMINATED

With all the technology and scientific discoveries, society could be transformed for the better. "By 2060, we could eliminate poverty and disease, but we have to want to do it. [Political] power and other things might not let us do it." Most likely, the majority will live in the Third World, but nanotech, for example, will be a good Third-World technology because it doesn't take infrastructure.

Bibliography

Ackerman, Forrest J. 1997. *Forrest J. Ackerman's World of Science Fiction*. Los Angeles: General Publishing Group.

Aldiss, Brian. 1975. *Science Fiction Art*. New York: Bounty Books.

Appelbaum, Stanley. 1977. *The New York World's Fair 1939/1940*. New York: Dover Publications.

Brosterman, Norman. 1999. *Out of Time: Designs for the Twentieth-Century Future*. New York: Harry N. Abrams.

Carroll, Robert Todd. 2003. *The Skeptic's Dictionary: A Collection of Strange Beliefs, Amusing Deceptions, and Dangerous Delusions*. New York: John Wiley & Sons.

Cerf, Christopher and Victor Navasky. 1984. *The Experts Speak: The Definitive Compendium of Authoritative Misinformation*. New York: Pantheon Books.

Clute, John. 1995. *Science Fiction: The Illustrated Encyclopedia*. New York: Dorling Kindersley Publishing.

Corn, Joseph J. and Brian Horrigan. 1984. *Yesterday's Tomorrows*. Baltimore, MD: Johns Hopkins University Press.

Crouse, William H. 1963. *Science Marvels of Tomorrow*. New York: McGraw-Hill Book.

Dewdney, A. K. 1997. *Yes, We Have No Neutrons*. New York: John Wiley & Sons.

Gardner, Martin. 2000. *Did Adam and Eve Have Navels?* New York: W. W. Norton & Co.

Gardner, Martin. 1952. *Fads and Fallacies in the Name of Science*. New York: Dover Publications.

Gardner, Martin. 1981. *Science: Good, Bad and Bogus*. Buffalo, NY: Prometheus Books.

Haber, Heinz. 1956. *The Walt Disney Story of Our Friend the Atom*. New York: Simon and Schuster.

Helfand, William H. 2002. *Quack, Quack, Quack: The Sellers of Nostrums in Prints, Posters, Ephemera & Books*. New York: The Grolier Club.

Hilton, Suzanne. 1978. *Here Today and Gone Tomorrow: The Story of World's Fairs and Expositions*. Philadelphia: Westminster Press.

Hyde, Margaret O. 1955. *Atoms Today & Tomorrow*. New York: McGraw-Hill Book.

Janicki, Edward. 1990. *Cars Detroit Never Built*. New York: Sterling Publishing.

Jay, Ricky. 1986. *Learned Pigs & Fireproof Women*. New York: Farrar, Straus and Giroux.

Kohn, Alexander. 1986. *False Prophets*. Oxford, UK: Basil Blackwell.

Kyle, David. 1976. *A Pictorial History of Science Fiction*. London: Hamley Publishing Group.

Launius, Roger D. and Howard E. McCurdy. 2001. *Imagining Space*. San Francisco: Chronicle Books.

Lent, Constanin Paul. 1947. *Rocketry: Jets and Rockets*. New York: Pen-Ink Publishing.

Ley, Willy. 1944. *Rockets: The Future Beyond the Stratosphere*. New York: The Viking Press.

McCoy, Bob. 2000. *Quack! Tales of Medical Fraud*. Santa Monica, CA: Santa Monica Press.

Menzel, Donald H. 1953. *Flying Saucers*. Cambridge, MA: Harvard University Press.

Regis, Ed. 1990. *Great Mambo Chicken and the Transhuman Condition: Science Slightly Over the Edge.* Reading, MA: Addison-Wesley Publishing.

Robinson, Frank M. 1999. *Science Fiction of the 20th Century.* New York: Barnes and Noble Books.

Shermer, Michael. 2001. *The Borderlands of Science.* Oxford, UK: Oxford University Press.

Shermer, Michael. 2002. *The Skeptic Encyclopedia of Pseudoscience Vol I & II.* Santa Barbara, CA: ABC CLIO.

Snyder, Robert, ed. 1980. *Buckminster Fuller: An Autobiographical Monologue Scenario.* New York: St. Martin's Press.

Tenner, Edward. 1986. *Why Things Bite Back: Technology and the Revenge of Unintended Consequences.* New York: Vintage.

Topham, Sean. 2003. *Where's My Space Age?* Munich: Prestel.

Wanjek, Christopher. 2003. *Bad Medicine.* New York: John Wiley & Sons.

Welsome, Eileen. 1999. *The Plutonium Files: America's Secret MedicalExperiments in the Cold War.* New York: Dial Press.

Wieners, Brad and David Pescovitz. 1996. *Reality Check.* San Francisco: Hardwired.

Youngson, Robert M. and Ian Schott. 1996. *Medical Blunders.* New York: New York University Press.

Youngson, Robert M. 1998. *Scientific Blunders.* New York: Carroll & Graf.

Resources

Thanks to Bob McCoy at the Museum of Questionable Medical Devices for use of its old medical pamphlets and to the vast archives of historical oddities at Simple Sense of Superiority.

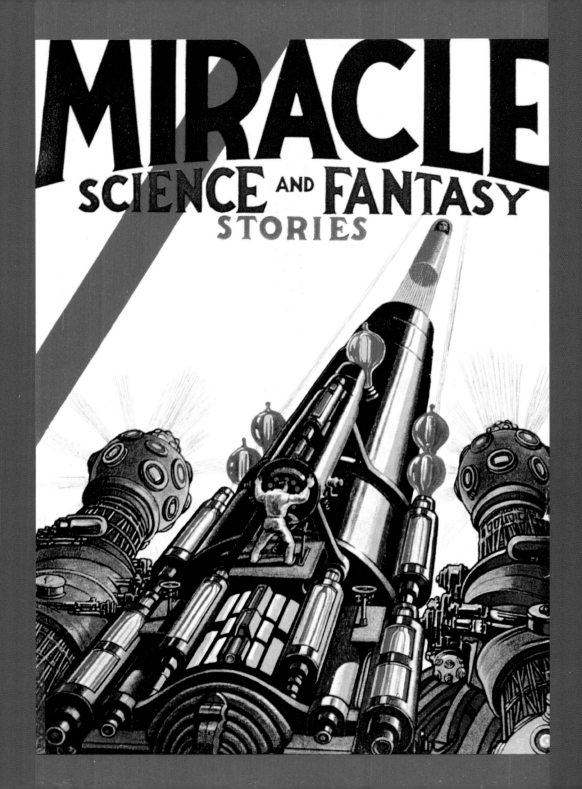